Ultraviolet LED Technology for Food Applications

Ultraviolet LED Technology for Food Applications

From Farms to Kitchens

Edited by

Tatiana Koutchma, PhD

Guelph Research and Development Center
Agriculture and AgriFood Canada
University of Guelph, Guelph, Ontario, Canada

ACADEMIC PRESS
An imprint of Elsevier

Academic Press is an imprint of Elsevier
125 London Wall, London EC2Y 5AS, United Kingdom
525 B Street, Suite 1650, San Diego, CA 92101, United States
50 Hampshire Street, 5th Floor, Cambridge, MA 02139, United States
The Boulevard, Langford Lane, Kidlington, Oxford OX5 1GB, United Kingdom

Ultraviolet LED Technology for Food Applications

ISBN: 978-0-12-817794-5

Publisher: Charlotte Cockle
Acquisition Editor: Nina Rosa Bandeira
Editorial Project Manager: Ruby Smith
Production Project Manager: Sreejith Viswanathan
Cover Designer: Greg Harris

Working together to grow libraries in developing countries

www.elsevier.com • www.bookaid.org

Contents

Contributors

Aušra Brazaitytė
Institute of Horticulture, Lithuanian Research Centre for Agriculture and Forestry, Lithuania

Pavelas Duchovskis
Professor, Lithuanian Research Centre for Agriculture and Forestry, LT-54333, Babtai, Lithuania

Andrew Green, Msc
Guelph Research and Development Center, Agriculture and Agri-Food Canada, Guelph, ON, Canada

Tatiana Koutchma, PhD
Research Scientist, Guelph Research and Development Center, Agriculture and Agri-Food Canada, Guelph, ON, Canada

Jurga Miliauskienė
Doctor, Lithuanian Research Centre for Agriculture and Forestry, LT-54333, Babtai, Lithuania

Vladimir Popović, MSc
Guelph Research and Development Center, Agriculture and Agri-Food Canada, Guelph, ON, Canada

Neringa Rasiukevičiūtė
Institute of Horticulture, Lithuanian Research Centre for Agriculture and Forestry, Lithuania

Giedrė Samuolienė
Doctor, Lithuanian Research Centre for Agriculture and Forestry, LT-54333, Babtai, Lithuania

Alma Valiuškaitė
Institute of Horticulture, Lithuanian Research Centre for Agriculture and Forestry, Lithuania

Viktorija Vaštakaitė-Kairienė
Institute of Horticulture, Lithuanian Research Centre for Agriculture and Forestry, Lithuania

Akvilė Viršilė
Doctor, Lithuanian Research Centre for Agriculture and Forestry, LT-54333, Babtai, Lithuania

Preface

*To my husband, son, friends, and colleagues who collaborated and gave me sup-
port and encouragement in my work at the first edition of this monograph*

As relatively new preservation and disinfection method, ultraviolet light (UV)
light has established positive consumer image and is of interest to the food industry
for processing of raw and prepared products, and sanitation of plant facilities. The
key drivers of UV technology have been its relative low cost compared with other
treatments, nonthermal, nonionizing, and nonchemical character, and dry, residue-
free, nontoxic processing nature. Contemporary advances in science and engineering of UV technology and profits for food industry applications have demonstrated
that UV holds considerable promise in food processing as an alternative treatment
for liquid products, ingredients, preprocessing method of raw materials or dry ingredients, postprocessing of ready-to-eat meals, shelf life extension method of fresh
produce, and dry sanitation method of no-food contact and food contact surfaces
at food plants. Also, UV light technology has been emerging as a harvest and postharvest method to improve efficiency, yield, quality of crops, and fresh produce. As
effective disinfection and sanitation technique, UV can be applied for numerous
point-of use applications at commercial and domestic kitchens. This means that
UV technology presents a new technological solution with enormous potential for
control of pathogens and forestalling spoilage throughout the food supply chain
from farm to kitchen. High-voltage, arc-discharge mercury or amalgam lamps that
can generate photons solely at 253.7 nm are commonly used and legally approved
for selected food application. The UV lamps have been historically proven to be
effective for municipal drinking water disinfection and, more recently, hospital surface sanitation. However, the UV lamps efficacy greatly depends on the species of
pathogen, the variety of produce, the initial degree of microbial contamination,
the topography of the treated surface, and where in production cycle the intervention
takes place. The uses of these UV light sources are generally not viewed as food
plant or food preparation area friendly. The equipment that houses them is bulky
and requires significant setup expenses and ongoing operational safety controls in
order to optimize germicidal light delivery, as well as, to provide requisite worker
protection. The solution has been found in exploring the emerging applications of
light-emitting diodes (LEDs) in food production. LEDs' performance has been
extensively studied and proven for low flow, point–of-use, drinking water treatment,
and curing applications and their effectiveness continues to be optimized from
manufacturing and applications point of view. The challenge is how to apply and
use numerous benefits of LEDs in food production chain for producing better crops
and produce in modern farms, postharvesting operations, sanitizing, and extending
shelf life of fresh products at retail and domestic levels.

Visible and UV LEDs are viewed as the upcoming solution for enhanced quality,
food safety, and reduced waste at different levels of supply chain. Also, LEDs can be

installed in areas where installation of current UV lamp equipment is problematic such as cold storage, product transport, and small, countertop, point of preparation, disinfection apparatuses. Particularly, UV LEDs are rapidly becoming more efficient and cheaper; hence it is expected that LED technology will become more attractive to the food industry in the near future.

In spite of available knowledge on particular aspects of LEDs' applications for curing, there is very limited information about LEDs in food production operations. No single source of information or monograph is currently available that would integrate modern fundamental and practical knowledge about LEDs light with current food applications and their challenges. Therefore, as the first edition of the book in the area of food production and processing "*Ultraviolet LEDs Technology for Food Applications from Farms to Kitchens*" will greatly benefit the food industry, academia, and researchers in the rapidly developing area of LEDs technologies. Research related to LEDs in the food industry mainly focuses on four areas of food production and horticulture, postharvest storage, food safety, and point-of-use applications. As more and more innovations take advantage of the unique properties of visible and UV LEDs, they will offer numerous novel and unique approaches to safety of food production. This book will focus on LED technology and LEDs' unique properties and advantages; summarize the developments and advancement in four areas of applications starting from produce production and horticulture, postharvest and postprocessing storage, safety, and point-of-use applications.

After a brief introduction in the LED technology and history of their development, the first chapter will review unique advantages of LEDs for foods and economical, energy saving, and sustainability aspects of LEDs applications for foods.

Chapter 2 discusses the features of manufacturing technology of LED light sources from chips to LED systems in visible and UV range. The focus of Chapters 3 and 4 is the review of current research and applications of LEDs in horticulture and crop production, postharvest preservation, and produce storage. In Chapter 5, the next steps of adaption of UV LED for safety applications by food industry are discussed. This chapter reviews the germicidal action of UVA, UVB, UVC, and blue light, existing research and first reported application of UV LEDs against foodborne pathogens at multiple wavelengths. Understanding of bacterial action spectra is also discussed to provide basis for optimization of the most effective treatment using single or multiple-wavelengths combinations. Considerations for UV LEDs treatment of fresh produce with some reported effects on quality and nutritional attributes are included.

The considerations of current technology state and summary of established and potential applications of LEDs in food production, processing, and safety conclude the first edition of the book.

Dr. Tatiana Koutchma is the editor of the book and the first author of Chapters 1, 2 and 5. Dr. Tatiana Koutchma is currently a research scientist in Novel Food Process Engineering, Agriculture and Agri-Food Canada. She has an extensive experience in emerging food processing technologies working with international government organizations, academia, and leading food companies to assist technology implementation and commercialization.

The first edition of the monograph *"Ultraviolet LEDs Technology for Food Applications from Farms to Kitchen"* is intended to provide LED manufacturers, food engineers, technologists and scientists, and undergraduate and graduate students working in research, development, and operations, with broad and readily accessible information on the available science and existing and potential applications of UV LEDs technology. With five chapters, this book represents the most comprehensive and ambitious undertaking on the subject of LED technology for foods that exists to date.

Tatiana Koutchma

Introduction

Alternative opportunities to current practices of food production and processing that are more sophisticated and diverse are being intensively investigated in the past decades. Ultraviolet light (UV) irradiation is of interest to the industry as a nonthermal nonionizing and nonchemical method of preservation, quality improvements, and risk mitigation. UV is currently used in food industry in a number of food and feed processing operations. This includes disinfection of facilities, decontamination of water and air, shelf life extension and safety enhancements of raw and finished products, sterilization of rinsing water, and enhancement of nutritional attributes through formation of vitamin D in yeasts, bread, milk, and fresh produce. UV light treatment has attracted attention and getting more interest in commercialization because UV light technology is friendly for environment; it does not produce drug resistance of bacteria without addition of chemical substances into treated object.

Conventional UV treatment uses UV lamps such as the low- or medium-pressure mercury and amalgam lamps. They emit UV of 253.7 nm in wavelength and this wavelength is broadly accepted that it has the highest antimicrobial effect and classified as germicidal UVC light. However, UV light is characterized by three regions such as UVA (320–400 nm), UVB (280–320 nm), and UVC (200–280 nm). The risk of UVA light photons is lower than light in the UVC range: it is not very hazardous for human eyes and skin than mercury lamps.

Recently, light-emitting diodes (LEDs), semiconductor diodes capable of producing light through electroluminescence, started emerging as alternative to UV lamps. Visible (blue, green, and white) and infrared LEDs became commonplace in many electronics and curing and lighting applications because of extremely high efficiency (up to 80%) and long lifetime around 100,000 h. The germicidal UV LEDs operate at wavelengths between 210 and 400 nm in three regions UVA, UVB, and UVC. Unlike traditional lamp light sources, whose output wavelength is fixed, UV LEDs can be manufactured to operate at the optimum wavelength or their combination for the application. Furthermore, mercury is not used in UV LED and it does not have harmful effects on ether a human body or environment. UVLED is a smaller device and may have longer operating life as compared with those of the low-pressure mercury lamps. UV LEDs will become a low power consumptive and environmentally friendly light source.

History of Visible and UV LEDs Development

German, British, and American researchers working in parallel to advance the electronics, phone, TV, and lighting industries developed first light-emitting diodes (LEDs) in the 1950s. The early LEDs used a gallium phosphide (GaP) substrate or doped in nitrogen or zinc oxide and emitted visible light in the red to green ranges. The blue LED proved indefinable for three decades. In the 1980s, three Japanese researchers developed the high-output blue LED by growing high-quality crystals into a multilayer gallium nitride (GaN)—based substrate. Isamu Akasaki, Hiroshi Amano, and Shuji Nakamura won the Nobel Prize in Physics in 2014 (Scientific Background on the Nobel Prize in Physics 2014) for this development and its impact on the lighting and electronics industries. Blue LEDs generate white light either by combining with red and green LEDs or by illuminating a phosphor coating. The breakthrough of the white LEDs transformed the electronics industry with the advent of computer screens and smartphones. White LEDs are used in lightning and surpassed incandescent bulbs because of almost 80% energy efficiencies and lifetime of 100,000 h.

Blue light LED overlaid the way for the ultraviolet LED (UV LED), which was developed through several iterations of manufacturing and material improvements. UV LEDs are constructed using multilayer substrates from indium gallium nitride, diamond, boron nitride, aluminum nitride, and aluminum gallium indium nitride and capable of emitting UV light at wavelengths as low as 210 nm.

First generation of UV LEDs were UVA LEDs in 320—398 nm range and were used to degrade harmful organic pollutants such as pharmaceuticals, insecticides, and dyes with the addition of photocatalysts. Also, coatings such as adhesives and inks have been cured using UVA LED technology.

Currently, UV LED technology is quickly extending to the UVB and UVC range as well. To be effective as traditional UV lamps in germicidal range, UVB and UVC LEDs must achieve higher wall plug efficiencies (WPEs). Given the current state of technology, the WPE of UVB and UVC LEDs is approximately 4%. However, UVB and UVC LEDs are on growing trend and on speed with the development of red, blue, and UVA LEDs. UVC LEDs are already being incorporated into point-of-use units to serve the defense and outdoor industries. They are used in airplanes to disinfect air in the passenger cabin, and they are being tested for water treatment.

Potentially, UV LEDs can be used for the treatment of beverages, disinfection of food surfaces, packaging, and other food contact and noncontact surfaces. Also, they can be installed in areas where installation of current UV lamp equipment is problematic such as cold storage, product transport, and small countertop, point of preparation, disinfection apparatuses and commercial and domestic refrigerators. In addition, UV LEDs are rapidly becoming more efficient and cheaper, hence it is expected that LED technology will become more attractive to the food industry in the near future.

Research related to LEDs in the food industry mainly focuses on four different areas of food production and horticulture, postharvest storage, food safety, and point-of-use applications. As more and more innovations take advantage of the unique properties of UVC LEDs, they will offer numerous novel and unique approaches to safety of food production. Also, UVC LEDs will emerge as environmentally friendly solution to save energy, lower reliance on toxic chemicals, improve worker and consumer safety, extend shelf life from field to kitchens, and improve processing.

This is the first monograph with a sole focus on UV LED technology and LEDs' unique properties and advantages. It summarizes the developments and advancement in LED technology and four areas of food-related applications starting from produce production and horticulture, postharvest and postprocessing storage, safety, and point-of-use applications.

Tatiana Koutchma, PhD
Guelph Research and Development Center
Agriculture and Agri-Food Canada
Guelph, ON N1G 5C9, Canada

Overview of Ultraviolet (UV) LEDs Technology for Applications in Food Production

1

Tatiana Koutchma, PhD [1], Vladimir Popović, Msc [2], Andrew Green, Msc [2]

Research Scientist, Guelph Research and Development Center, Agriculture and Agri-Food Canada, Guelph, ON, Canada[1]; Guelph Research and Development Center, Agriculture and Agri-Food Canada, Guelph, ON, Canada[2]

Chapter outline

Ultraviolet LED Technology for Food Applications. https://doi.org/10.1016/B978-0-12-817794-5.00001-7

Abstract

Ultraviolet light (UV, UVC) disinfection is a well-established technology for air, water, and surfaces treatments. In the last decade, UV technology found more applications in food production chain because this is economic, effective, and versatile dry processing technology that can potentially improve safety and preservation quality of different categories of foods.

Traditional way of generating UVC light is using high-voltage arc-discharge mercury or amalgam lamps that can generate photons solely at 253.7 nm. Light-emitting diodes or LEDs are semiconductor devices that can also emit UVC light, but depending on the material properties of the diodes, they can emit photons at multiple wavelengths in the UV range between 255 and 365 nm. The germicidal effects of UV LEDs against bacteria, viruses, and fungi already have been demonstrated and reported along with the first applications for disinfection of air, water, and surfaces made for the "point of use" integration. Despite the fact that there are no commercial applications in food production yet, the UV LEDs are the next wave in the LED revolution that can bring numerous advantages for food processing, safety, and plants facilities disinfection.

This chapter will present the introduction in basics of LEDs technology and will discuss the unique advantages of this technology for food safety applications. Also, economical, energy saving, and sustainability aspects of UV LED applications are introduced.

Keywords: Food applications; Safety; UV light; UVC light-emitting diodes (UVC LEDs).

UV Light as an Emerging Food Safety Technology

For many years, ultraviolet (UV) light has been considered as a technology that suits solely for water and waterlike UV transparent fluids because of the challenges of low penetration in foods and beverages. The first recorded use of UV light as a means of disinfection occurred in the late 1800s (Downes & Blunt, 1878). In these early systems, the sun was used as a light source to disinfect water. In 1903, the Nobel Prize for medicine was awarded to Niels Finsen, who discovered that UV light could be used to inactivate pathogenic organisms (The Nobel Foundation, 1903). By 1910, the first UV water disinfection unit using early mercury vapor lamps (Hewitt lamps) was deployed in France (U.S. Army Public Health Command, 2014). The development of lamps capable of producing UV photons furthered the advancement of UV-mediated disinfection technologies by removing reliance on the sun. The tubular low-pressure mercury (LPM) lamps developed by General Electric in the 1930s led to wider use of the technology both for germicidal purposes and for general lighting.

Recently, following the successful applications in water and juice treatments that demonstrated the high inactivation efficiency of UVC light, the UV technology started to emerge and became one of the most promising nonthermal and nonchemical preservation processes that slowly being adopted in food processing. Industrial

UV systems have been developed for new food and beverages applications that are capable to deliver efficient processes targeted against pathogenic and spoilage contaminants in many categories of fluid food products, drinks, and ingredients. The examples of existing and potential applications of UV light include fruit and vegetable juices, teas and coffees, milk, sugar syrups, liquid eggs and egg components, wine, and protein ingredients.

Given the expectations of minimum effects on products quality, flavor, nutritive content, and overall health benefits, the use of nonthermal, nonionizing, and nonchemical technologies such as UV light is considered as more advanced, less expensive alternative processing option for foods. The advantages associated with UV treatment are that it is effective against food, water, and airborne pathogens and can control pathogens level in food processing and storage facilities. No known toxic or significant nontoxic by-products are formed during the treatment, certain organic contaminants can be removed, no off taste or odor is formed when treating foods and beverages, and the treatment requires less energy when compared with thermal pasteurization processes. The US FDA approved UVC for treatment of juice products to reduce human pathogens and other microorganisms (US FDA, 2001), processing water, and surfaces.

Additionally, with the growth of fresh, natural, and ready-to-eat markets, UV and other light technologies are used as a dry food safety and preservation technique when applied for treatment of the food products surface because of their unique antibacterial effects and minimal impact on food quality. The effective reduction of microbial contamination on the surfaces of whole fruits and vegetables, fresh-cut produce, raw meat and sea food products, eggs, sliced cheese, flour powder, and spices have been demonstrated in numerous reports (Koutchma, 2014).

UV light treatments are emerging economical physical interventions toward improved hygiene control measures in the food industry. Sanitation, decontamination, disinfection, and oxidation with UV light is a versatile, environmentally friendly technology, which can be used in the food processing, transportation, and storage facilities to reduce microbial contamination of food contact surfaces and consequently to improve safety of finished products.

Continuous UV Light Basics and Germicidal Effects

Ultraviolet (UV) light is any light on the electromagnetic spectrum with a wavelength between 100 and 400 nm (Koutchma, Forney, & Moraru, 2009). UV light is typically subdivided into four categories: long wave UVA, with wavelengths of 315—400 nm; middle wave UVB, with wavelengths of 280—315 nm; short wave UVC, with wavelengths between 180 and 280 nm; and vacuum UV (VUV), with wavelengths of 100—180 nm (Fig. 1.1).

UVA range is considered to be responsible for changes in human skin or tanning; UVB range can cause skin burning and possibly lead to skin cancer; and UVC range from 200 to 280 nm is considered the germicidal range because it effectively

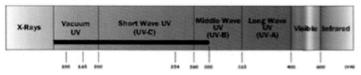

FIGURE 1.1

The UV portion of the electromagnetic spectrum.

Modified from https://www.canada.ca/en/health-canada/services/sun-safety/what-is-ultraviolet-radiation.html.

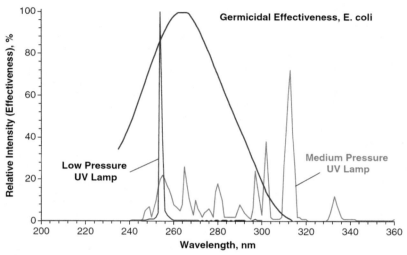

FIGURE 1.2

Cell deactivation depending on the exposed wavelength of UV light and operated wavelength of mercury lamp.

From https://commons.wikimedia.org/wiki/File:Germicidal_Effectiveness_for_LP_%26_MP_mercury_lamp.
png#mw-jump-to-license.

inactivates bacterial, viral, and protozoan microorganisms because of the fact that microbial DNA absorbs UV light most strongly at approximately 260 nm (Koutchma et al., 2009) as shown in Fig. 1.2.

It has been shown that UVC light inactivates microorganisms primarily by inducing the formation of pyrimidine dimers in DNA (Goodsell, 2001). Light is absorbed by the double or triple hydrogen bonds keeping adenine-thymine and guanine-cytosine base pairs (respectively) together in the DNA strand. When this occurs in two pyrimidine bases adjacent to each other in the DNA strand, there is a chance for a cyclobutane dimer or a 6,4-pyrimidine-pyrimidone dimer to form between them, the former being more common, the latter being more mutagenic. These photoproducts cause a lesion in the DNA, which inhibits DNA and RNA polymerases, in turn inhibiting gene replication and expression, leading to cell death (Jaspers, 2001).

It has been believed that the wavelength of 253.7 nm is most efficient in terms of germicidal effect because photons are most absorbed by the DNA of microorganisms at this specific wavelength. Light with a wavelength below 230 nm is most effective for the dissociation of chemical compounds.

However, UVB wavelengths have also been shown to be capable of inactivating bacteria, through similar mechanisms to that of UVC (Sinton, Hall, Lynch, & Davies-Colley, 2002). UVB light is also thought to inactivate bacteria partially by damage to other cellular structures in addition to DNA (Li, Wang, Huo, Lu, & Hu, 2017). It has been demonstrated that UVA light causes germicidal effects at high doses (Li et al., 2010; Mori et al., 2007; Nakahashi et al., 2014). UVA light causes damage to a number of cellular structures via the formation of reactive oxygen species from the photooxidation of oxygen and/or water or of the cellular structures themselves. This in turn causes oxidative damage to structures such as membrane lipids, proteins, and DNA (World Health Organization, 1994).

Sources of Continuous UV Light
Low-Pressure Mercury Lamps

Traditionally, UV light is emitted by the inert gas flash lamp that converts high power electricity to high power radiation. A few types of continuous light UV lamps are commercially available that include low- and medium-pressure mercury lamps (LPM and MPM), low-pressure amalgam (LPA), and excimer lamps (ELs). LPM and MPM lamps are the dominant sources for UV light treatment of fluid foods, drinks, and beverages including water processing. However, only LPM lamps that solely emit UV light at 253.7 nm are currently approved by the US FDA for food applications (US FDA, 2001).

A substantial body of evidence has been generated to demonstrate the efficacy of LPM lamps on a variety of surfaces (Andersen et al., 2006; Kowalski, 2009; Rutala et al., 2010), as well as in water (Hijnen et al., 2006) and in air. UVC light has been shown to be effective against pathogens and spoilage organisms on the surfaces of a diverse selection of fresh produce and processed foods.

Low-pressure mercury lamps (LPM) are typically used as sources to produce UVC light via the vaporization and ionization of mercury. This type of lamp consists of three electrodes—two main electrodes and one secondary starting electrode connected through a resistor to a main electrode (Fig. 1.3). These electrodes are contained within a tube made of quartz (or other UV-transparent material) filled with argon gas and a small amount of liquid mercury. When the lamp is turned on, the applied voltage results in the production of an arc between the starting electrode and the adjacent main electrode. This initiates the formation of an arc between the two main electrodes (on opposite ends of the tube), which produces enough heat to vaporize and subsequently ionize the liquid mercury within the tube

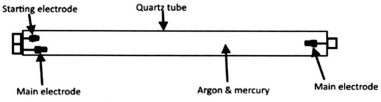

FIGURE 1.3

Schematic diagram of a low-pressure mercury lamp.

(The Mercury Vapor Lamp, 2016). The ionized mercury vapor emits several narrow bands of light including two bands in the UVC range: 253.7 and 184 nm, latter of which is strongly absorbed by the quartz tube as well as the atmosphere. However, this wavelength can be made useable when lamps are outfitted with a synthetic quartz tube (which does not absorb the 184 nm band as strongly) and operated in a vacuum.

Lamp manufacturers use the following characteristics to compare electrical and germicidal efficiency of UV sources: total power input, UVC efficiency, irradiance, and irradiance at a given distance and lamp lifetime.

1. Total power input (W) depends on voltage and electric current.
2. UVC efficiency of the lamp is evaluated based on the ratio of measured wattage of UV output in the spectral range of interest versus the total wattage input to the lamp.
3. Irradiance is the amount of flux incident on a predefined surface area and most commonly expressed in $mW\ cm^{-2}$ or $W\ m^{-2}$.
4. UV irradiance at a given distance from the lamp(s) surface is 10 cm or 20 cm.
5. Lifetime is a total number of hours of UV source operation when emitted light irradiance is higher than 80%–85% of initial output value.

Typically, technical characteristic of LPM UV source serve as a basis for performance comparison with other continuous UV light sources. Although LPM lamps have high power output and germicidal efficacy, they also have several disadvantages in food processing applications:

1. Contain toxic mercury
2. Fragile (contain glass)
3. Power output decreases significantly at low temperatures in LPM lamps
4. Limited lifetimes (maximum18,000 h)
5. Require warm-up time to reach maximal irradiance
6. Bulky size limits operational integration

Recently, advancement in light-emitting diode (LED) technology has allowed for the development of UV light sources that can circumvent the above noted disadvantages while offering several other new advantages.

UV LEDs

An LED is a semiconductor diode capable of producing light through electroluminescence. LEDs are a type of semiconductor consisting of p-n junction—a junction between a p-type and n-type semiconducting material; the p-type being a material that is electron-poor (i.e., has a high concentration of electron holes), and the n-type being a material that is electron-rich (Fig. 1.4). When current is passed across the diode, electrons from the n-type material are able to combine with electron holes in the p-type semiconductor (occurring in what is sometimes referred to as an active layer), causing excited electrons to release their energy as photons (Schubert, 2006). The wavelength of these photons is determined by the bandgap energy (the quantity of energy required to promote an electron from the valence band to the conduction band of the material) of the semiconducting material (Schubert, 2006).

Visible (400–800 nm, blue, green, and white) and infrared (800–1000 nm) LEDs became commonplace in many electronics, curing, and lighting applications because of extremely high efficiency (up to 80%) and long lifetime around 100,000 h.

UV LEDs are light-emitting diodes that emit photons with wavelengths in the UV range (100–400 nm). Typically, UV LEDs are made from either aluminum nitride (AlN) or aluminum gallium nitride (AlGaN). These conductors are differentially doped (intentionally made to have impurities) to produce the aforementioned electron holes (for p-type) or electron-rich regions (for n-type). Altering the type of dopant will alter the bandgap of the material, and therefore the energy of the photons emitted.

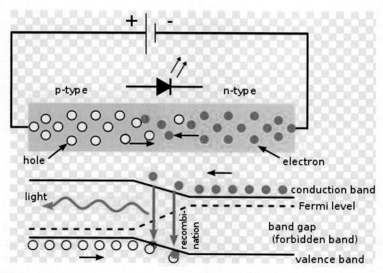

FIGURE 1.4

Schematic diagram of a light-emitting diode.

From https://en.wikipedia.org/wiki/Light-emitting_diode_physics#/media/File:PnJunction-LED-E.svg.

New advances in semiconductor materials, such as gallium nitride (GaN) and aluminum nitride (AlN), have made it possible for LEDs to emit light into the UV range. This breakthrough was of critical importance for the further implementation of UV light technology in food safety applications.

Currently, UV LEDs are available in a wide variety of discrete wavelengths from 210 up to 400 nm, including many wavelengths in the germicidal UVC region. UV LEDs are commercially used with increasing frequency in applications such as curing of inks and adhesives (which often requires UVA light) as UVA LEDs can be made with significantly higher power outputs compared with UVC and UVB LEDs (Forte, 2014).

Other applications include use of UVA light for body fluid detection in forensics, protein analysis, and the detection of counterfeit bills. In general, UV LEDs have a major advantage over other UV light sources because of their small size, low power draw, and their ability to emit a wide variety of discrete UV wavelengths.

Comparison of Technical Specifications and Performance of UV LEDs with Low-Pressure Lamps

As LPM lamps are considered the benchmark in UV disinfection technology for foods, comparison between UV LEDs and LPM lamps is critical for assessing the feasibility of UV LEDS as emerging UV technology. Table 1.1 summarizes technical specifications and performance characteristics of both sources (Table 1.1).

Although many studies have made such comparisons, it is important to note that a true comparison (i.e., under identical treatment conditions) of these two UV light sources is likely not possible because UV LEDs have wider emission peaks than LP lamps, which could impact germicidal efficacy as shown in Fig. 1.5. Furthermore, manufacturing of UV LEDs that emit light at the same wavelength as LPM lamps (253.7 nm) is still challenging. Therefore, comparisons are often made at different wavelengths while keeping all other treatment conditions constant (i.e., UV dose, bacterial strain, treatment media or surface, etc.).

Wavelength

The wavelength emitted by an LED is defined by the material properties of the epilayers. The advantage of UVC LEDs is that they can be designed to produce continuous UVC light at the optimal germicidal wavelength tuned against specific microorganisms. As shown in Table 1.1, UV LED can cover range from 210 up to 395 nm in UVA, UVB, and UVC diapasons as opposed to LPM lamps, which are limited to emission at 253.7 nm. The shortest wavelength LED produced consisted almost entirely of aluminum nitride and emitted at 210 nm. This, however, was only a proof of concept and emitted very little light. Currently, the nature of the dopants used to produce carriers in AlN and AlGaN is such that the wavelength limit of commercial UV LEDs is around 235 nm.

Table 1.1 Comparison of Technical and Performance Characteristics Between Traditional Low Pressure Lamps and UV LEDs.

Parameter	Low-Pressure Lamps	UVC/UVB LEDs	UVA LEDs
Wavelength (nm)	253.7	210–320	320–398
Source material	Mercury or mercury amalgam	GaN or AlN	GaN or AlN
Microbial inactivation treatment times	Seconds to minutes	Minutes to hours	Hours
Optical power output	5–80 W[a] 30–600 W[b]	0.1–100 mW	0.1–5 W
Electrical to germicidal efficiency (%)	30–40	0.2–9.5	NA
Lifetime (h)	8,000–18,000	1,000–10,000	30,000
Optimal operational temperature (°C)	20–45[a] 0–60[b]	0–45	0–45
Lamp surface temperature (°C)	40[a] 90–120[b]	Ambient	Ambient
On/off	Minutes	Instant	Instant
Durability	Fragile	Rugged	Rugged

[a] *Mercury lamps.*
[b] *Mercury amalgam lamps.*

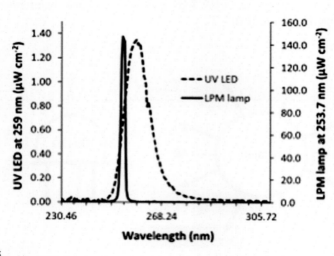

FIGURE 1.5

Peak UV light emission spectra of a UV LED emitting at 259 nm and a traditional LP mercury lamp emitting at 253.7 nm.

The following multiple considerations should be considered to choose UV LEDs wavelength to achieve optimal performance depending on application:

- UV LED cost that varies by wavelength. Rapid advances and growth in UVC LED manufacturing and increasing volumes are steadily driving down prices.
- UV LED lifetime. There is a high degree of variability between LED chips manufacturers' specifications related to lifetime at different wavelength values.
- UV LEDs optical output. Output power of a UVC LED device varies largely based on wavelength.
- Action spectra of target organisms. Pathogens and indicator organisms have different spectral sensitivity for different UV wavelengths.
- UV transmittance (UVT) of treated product. The UVT of the fluid being treated is a critical parameter for sizing UV systems and choosing a correct UV source. Different UVT values and absorbance characteristics must be accounted for optimal LED performance.

The most effective solution often lies at the intersection of many of the above factors and the most effective germicidal wavelength is not always the correct application wavelength.

Thermal Energy and Operating Temperature

The surface temperature of UV LEDs, LPM, and LPA lamps can differ significantly. Unlike LPM lamps, the light-emitting surface of UV LEDs does not expel thermal energy in the same direction as UV light photon energy and therefore remains at ambient temperature (Fig. 1.6).

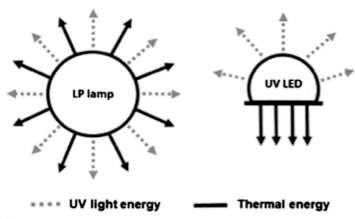

···· UV light energy   **━━ Thermal energy**

FIGURE 1.6

Difference in the emission of UV light and thermal energy from low-pressure lamps and UV LEDs.

In addition, the effect of ambient temperature on optical power output differs among low-pressure lamps and UV LEDs. The optical power output of mercury lamps has been shown to decrease significantly at refrigeration temperatures and therefore has a narrow operational temperature range (Shin, Kim, Kim, & Kang, 2015) (Table 1.1). UV LEDs, on the other hand, show a minor increase in optical power output as operating temperature approaches 0°C. This means that UV LEDs have operational advantage at refrigerated temperatures. LPA lamps can be operated successfully in a wide temperature range, but cooling is required to prevent excessive heating of food products because of high lamp surface temperatures around 90°C.

Power Output

One of the key differences between UV LEDs and LP lamps is their optical power output (or radiant flux), which is used to define the rate of flow of radiant energy per unit time and is expressed in watts (W) (Table 1.1). This parameter dictates the maximal UV irradiance and, in turn, the treatment time required for a light source to achieve a particular UV dose. The difference in optical power output between LPM lamps and UV LEDs is ultimately determined by the wall plug efficiency (WPE) of the light source. Wall plug efficiency (or radiant efficiency) is defined as the ratio of optical power output to input electrical power. Also, the efficiency of LPM lamps is often described by germicidal (or UVC) efficiency because the output optical energy is not fully emitted in the UVC range. Currently, the WPE of most commercial UVC LEDs is between 0.2% and 5%, which is well below that of LPM lamps (30%–40%), UVA LEDs (40%–60%), and visible LEDs (60%–81%). However, commercial UVC LEDs with WPE between 5% and 10% (SETi, 2012) have been manufactured, whereas researchers are exploring UVC LEDs that could exceed 40% WPE (Kheyrandish, Mohseni, & Taghipour, 2018). It is estimated that for every decade of UV LED development, optical power output will increase 20-fold, whereas price will decrease 10-fold. The optical power output of single UVC LED chips has already reached as high as 100 mW. Despite this, in a worst-case scenario, LPM and LPA lamps are currently able to generate anywhere between 50- and 6000-fold higher optical power values when compared with UVC and UVB LEDs (Table 1.1). This has hindered the use of these LEDs for food safety applications because of the overall higher UV doses and consequently treatment times required for microbial inactivation when compared with clear liquids and smooth artificial surfaces. For this reason, the majority of research and applications within the UVB and UVC LED segment to date has focused on the disinfection and purification of water and smooth surfaces. Unlike UVC and UVB, the irradiance values reported for UVA LEDs (up to 5 W) are comparable with those of LPM lamps. Despite this, UVA light has shown very little germicidal efficacy in comparison with UVC and UVB light and is therefore rarely used for disinfection purposes.

Fig. 1.7 shows previously reported UV irradiance values (standardized per single UV light unit at a distance of 1 cm) for LPM mercury lamps and UV LEDs at various

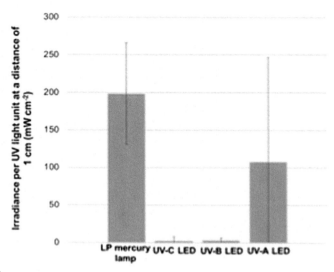

FIGURE 1.7

Comparison of average UV irradiance values for LPM lamps and UV LEDs standardized at a treatment distance of 1 cm and emitting from a single light source (lamp or LED chip). Values were compiled from previously published research studies.

wavelengths. These data show that, on average, single UVC and UVB LED chips produce significantly lower power output than a single LPM mercury lamp. However, the difference in power output could be slightly exaggerated owing to the discrepancy in size between the two light sources. The consequence of low power output is also evident in the small average treatment distance (5.9 cm) between UV LEDs and samples used in the tests to obtain sufficient UV doses (personal communication data). This is a concern because it has previously been shown that light emitted by UV LEDs requires at least 12.5 cm for proper collimation.

Consequently, UV LEDs lower power output and irradiance values result in longer exposure times (minutes and hours) to deliver sufficient UV dose to reach targeted inactivation.

However, the advantage that UV LEDs have over LPM lamps in terms of power output is the ability to generate higher power densities because of their compact size. In other words, UV LEDs can deliver more photons per surface area than traditional LPM lamps or even medium-pressure lamps. This trail is utilized for UVA LED curing applications.

Lifetime

Another parameter differentiating LPM lamps and UV LEDs is their lifetime. Lamp lifetime) is defined as the time required for UV light output to reach 80%−85% of its original value. Although they have made strides in recent years, UVC LEDs have shorter lifetimes compared with LPM lamps. UVC and UVB LEDs lifetime is approximately between 1000 and 10,000 h to reach 70% initial power output for UVC LEDs emitting at 255 and 275 nm, respectively, compared with approximately of 12,000—18,000 h for LPM lamp.

However, the operational lifetime of UV LEDs can be extended because of their instant on/off capabilities (i.e., pulsing) as they do not have to be continuously operated. Low-pressure lamps not only require a warm-up time but also suffer from a loss of power output with each on/off cycle. This means that they often remain operational for the entirety of the UV treatment process to prolong their lifetime. The lifetime of UVA LEDs has surpassed even LPM lamps approaching almost 30,000 h that is close to visible LEDs.

UV LEDs Market Development

The global UV LED market has been overtaking the traditional UV light market and has grown from $45 million USD in 2012 to $269 million in 2017 and is expected to reach $1.1 billion by 2023. The market is currently dominated by UA LEDs, which are mostly used for curing applications. Industrial implementation of UVA LEDs has outpaced UVB and UVC LEDs because of the fact that current manufacturing costs are reduced at increasing wavelength, whereas power output and wall plug efficiency (WPE) are increased. However, the disinfection and purification segment, which utilizes the far more antimicrobial effective UVB and UVC LEDs, is expected to achieve the highest growth in the next several years as WPE and power output of these LEDs continue to increase, whereas price decreases. Currently, the WPE of most commercial UVC LEDs is between 0.2% and 5%, which is well below that of LP lamps (30%—40%). However, several commercial manufacturers have recently produced UVC LEDs with WPE between 5% and 10%, whereas researchers are exploring UV LEDs that could exceed 40% WPE. The power output of single UVC LED chips has also reached as high as 100 mW. Furthermore, it is estimated that for every decade of UV LED development, power output will increase 20-fold, whereas price will decrease 10-fold. Until recently, the majority of research within the UVB and UVC LED segment has been focused on the disinfection and purification of air, water, smooth contact surfaces, and packaging.

Unique Advantages of UV LEDs for Foods From Farm to Kitchens

Despite there are no commercial food applications yet except for packaging, the UVC LEDs are the next wave in the LED revolution that can bring the numerous advantages of UVC disinfection for food processing, improvement of quality preservation, and food plant safety. The germicidal effects of UVC LEDs against bacteria, spores, viruses, and fungi already have been demonstrated and reported along with the first applications for disinfection of air, water, and surface made for the "point of use" integration. Germicidal LEDs at multiple wavelengths hold great promise for advancing food safety. UVC LEDs-based light fixtures will become the driving force behind wider adoption. Potentially, UVA, UVB, and UVC LEDs can be used for the treatment of beverages, disinfection of food surfaces, packaging, and other food contact and noncontact surfaces. Also, they can be installed in areas where installation of current UV lamp equipment is problematic such as cold storage, product transport, and small, countertop, point of preparation, disinfection apparatuses.

Here are some unique advantages of UVC LEDs for food applications.

Improved Inactivation Efficacy at Optimal UVC Wavelength

UV LEDs can be made to generate continuous light at the optimal germicidal (UVC) wavelength tuned against specific microorganisms. The ongoing research showed the effectiveness of UVC LEDs emitting in 255–365 nm diapason against common foodborne pathogens such as *Escherichia coli*, *Salmonella*, and *Listeria monocytogenes* (Green et al., 2018). The UVC LEDs emitting light in the approximate range of 265 and 280 nm were found to have the greatest efficacy in terms of bacterial inactivation and overall power output. Also, it was reported that UVA LEDs at 365 nm and/or near UV wavelengths (405, 460, 520 nm) have shown to reduce *Salmonella* on fruits.

Furthermore, these results can be used to identify the optimum inactivation wavelength for common food pathogens and hence increase processing efficiency. In some cases, germicidal efficacy can be improved by combining different UV wavelengths to produce a synergistic inactivation effect and tuned by design to match the most effective inactivation wavelengths in a given environment.

Shelf Life Extension

Perhaps the most widespread application for UV LEDs is general lighting. According to research from the Kansas State University, a switch to LED lights in refrigeration units can save retail meat industry million of dollars. Using LEDs allows to save energy and extend the color shelf life of some beef products. Five meat products displayed under LEDs lighting had colder internal product temperature compared with fluorescent light that helped to extend their shelf life. Better wavelength match can result in higher system efficiency and thus, additional energy savings. The list of UV LEDs applications in meat and poultry industry will certainly grow.

UVC LEDs can become a promising new UV source in the fresh produce industry because of risks associated with food poisoning and simultaneously extending produce storability and minimizing losses. For example, USDA scientists tested UVC LEDs in the range of 285–305 nm to extend shelf life of fresh fruits and vegetables in domestic refrigerators. Shelf life increase of $2\times$ was achieved using 20 mW m^{-2} of UVC LEDs power.

The results of the ongoing research in the Agriculture and Agri-Food Canada (AAFC) have shown that 277 nm UVC LEDs are capable of reducing mold spores on the skin of apples by 2.8 log, compared with 2 log at 253.7 nm using LPM lamp at an equivalent dose of 500 mJ cm^{-2}.

Operating Advantages

Being miniature, robust, and operating with a low electrical power, UVC LEDs can be manufactured with a highly stable output, operating at the optimum wavelength for the application. In addition, LEDs are mercury free, with no warm-up time, and potential long lifetime that can make them ideal for a variety of processing solutions. Also, UVC LEDs can be used in cold storage environments as they have shown to increase power output as temperatures approach 0°C, which further increases their versatility in terms of apparatus design. Green et al. (2018) showed the reduction in the UV treatment times by 6 s for *L. monocytogenes* and by 13 s for *L. seeligeri* because the UV LEDs exhibited an increase in irradiance E0 at 4°C (Table 1.2).

Previous studies by Shin et al. (2015) have also shown that although UV LED irradiance increased at colder temperatures, the germicidal efficacy against *L. monocytogenes* did not change significantly following 1 min of UV irradiation at 275 nm at 0 and 37°C on selective media.

Flexibility of Design or Point of Use Applications for Bench-Top Experiments, Processing, and Sanitizing Solutions

Multiple-wavelength UV LED chips have great potential for use in bench-top scale validation experiments. Their small size means that they can be easily integrated into a collimated beam apparatus for the purpose of studying inactivation

Table 1.2 Comparison of *L. monocytogenes* and *L. seeligeri* Reduction and UV Treatment Parameters for at Room (25°C) and Refrigeration (4°C) Temperatures Using a 268 nm UV LED Unit With a UV Dose of 7 mJ·cm^{-2}.

Organism	25°C			4°C		
	E0 (mW cm^{-2})	Treatment Time (s)	LCR	E0 (mW cm^{-2})	Treatment Time (s)	LCR
L. monocytogenes	0.138	76	4.68 ± 0.13	0.149	70	4.64 ± 0.26
L. seeligeri	0.138	80	3.56 ± 0.16	0.149	73	3.18 ± 0.31

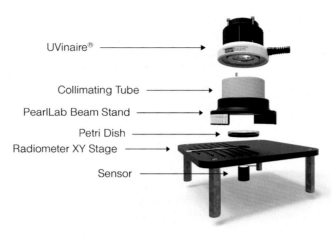

UVinaire® ——————————→

Collimating Tube ——————————→

PearlLab Beam Stand ——————————→

Petri Dish ——————————→

Radiometer XY Stage ——————————→

Sensor ——————————→

Petri dish and sensor not included

FIGURE 1.8

UV LEDs-collimated beam unit.

Used with permission from AquiSense Technologies, KY, USA.

kinetics at multiple wavelengths and wavelength optimization on surfaces and in liquid solutions. The UVC LED-collimated beam unit (Fig. 1.8) can be used for UV treatments at 259, 268, 289, and 370 nm to select the optimal wavelengths or their combination against food pathogenic organisms and find appropriate nonpathogenic surrogate. This unit consisted of an array containing three LEDs of each wavelength with thermal management, collimated tube, stand, and petri dish and allows conducting tests in controlled conditions and measuring UV exposure.

The selected LED chips can then be used in customized devices to scale up a larger disinfection device by establishing the operating UV doses for specific product or produce application and characterizing intended technical effect. The example of UV disinfection box that allows treating the whole fresh produce sample is shown in Fig. 1.9. Also, it can be used for the purposes of the process validation and studies of the effects of product and process critical parameters on UV dose delivery.

UVC LEDs chips enable flexible modular design over rigid lamps and their required bulky ballasts. UVC LEDs can easily be incorporated into simple water filters providing a highly effective solution for potable water needs that is easy to install and use with minimal supervision, maintenance, and space. Also, UV LED modules can be installed in the areas where current UV lamps cannot be used such as cold storage facilities, transport, and small disinfection apparatuses. For example, the PearlAqua system (Fig. 1.10) is an inline water disinfection device that can be used practically anywhere because of its small size and low power draw.

FIGURE 1.9

Customized UV LED disinfection box.

Used with permission from AquiSense Technologies, KY, USA.

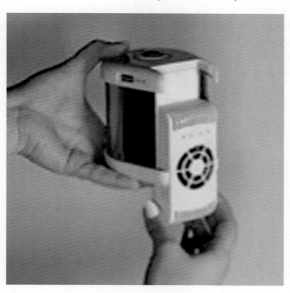

FIGURE 1.10

PearlAqua UV LED water disinfection system.

Used with permission from AquiSense Technologies, KY, USA.

UpSizing From Bench-Scale to Industrial Disinfection Applications

Within the next decade or so, UV LEDs will likely fill many of the roles that LPM lamps currently fill in terms of UV-mediated safety solutions for liquids and surfaces. Modular devices such as the Semray UV LED unit use UVA LEDs for industrial curing purposes (Fig. 1.11). Replacing these chips with UVC LEDs emitting a germicidal wavelength offers devices like this potential for use in the disinfection of food products moving along a conveyor as well as the conveyor itself.

There is also the potential to integrate germicidal UVC EDs into industrial water and beverage disinfection. Currently, UV light is one of the most popular disinfection methods for municipal drinking water and waste water. Replacing LPM lamps in these disinfection apparatuses with energy-efficient UV LEDs could result in significant energy savings for these municipalities, while eliminating the risk of lamp breakage leading to mercury contamination of the water. UV LEDs could also be integrated into water-assisted disinfection of solid food, where the food (usually fresh or minimally processed produce) is submerged in water under constant agitation and the entire rinse is exposed to UV light. UV LEDs could be used in place of LPM lamps in devices such as the turbulator, which uses UV light to disinfect produce submerged in water, agitated by air jets (Fig. 1.12).

UVC LED chips enable flexible modular design over rigid lamps and their required bulky ballasts. UVC LEDs can easily be incorporated into simple water filters, providing a highly effective solution for potable water needs that is easy to install and use with minimal supervision, maintenance, and space. Also, UV LED modules can be installed in the areas where current UV lamps cannot be used such as cold storage facilities, transport, and small disinfection apparatuses. The design rules for UV LED systems open new opportunities of what can be disinfected because UV sources are no longer limited to a long tube, but can mount the LEDs in flat panels, on flexible circuit boards, on the outside of cylinders so the options are almost endless.

FIGURE 1.11

UV LED curing unit.

FIGURE 1.12

Water-assisted UV disinfection apparatus.

Economical, Energy Saving, and Sustainability Aspects of LEDs for Foods

To displace the incumbent mercury-based technology, UV LEDs require further development to improve device efficiency and lifetime. The biggest challenge with UVC LEDs shorter than 265 nm is that they are still weak light emitters and have relatively short lifetime.

Currently, UVC LEDs are costly and emit at a relatively low irradiance. The relationship between power output and inactivation efficacy of UVC LEDs is an important consideration when selecting a UV treatment wavelength(s). For example, the 259 nm UVC LED had a lower irradiance ($E_0 = 0.0215$ mW·cm^{-2}) compared with the 268 nm UVC LED ($E_0 = 0.138$ mW cm^{-2}), resulting in significantly longer treatment times to achieve an equal UV dose, even though the germicidal efficacy at both wavelengths was similar.

When selecting a UV LED wavelength for industrial purposes, it is important to take into consideration the current trade-off in terms of germicidal efficiency and efficacy because their power output will significantly decrease with decreasing emission wavelength. This is the current disadvantage of UV LEDs compared with other UV sources. Therefore, UV LEDs emitting in the approximate range of 265–280 nm are recommended for use in industrial processes because of the fact

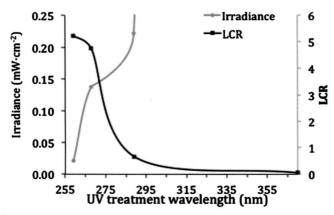

FIGURE 1.13

Relationship between UV LED wavelength, irradiance, and log count reduction (LCR) of
E. coli O157:H7 using UV LEDs between 259 and 370 nm and an equivalent UV dose of
7 mJ·cm^{-2}.

From Green, A., Popović, V., Pierscianowski, J., Biancaniello, M., Warriner, K., & Koutchma, T. (2018). Inac-
tivation of Escherichia coli, Listeria and Salmonella by single and multiple wavelength ultraviolet-light emitting
diodes. Innovative Food Science & Emerging Technologies, 47, 353–361.

that germicidal efficacy significantly decreases at wavelengths longer than 280 nm
offsetting any additional gains in power output. However, at wavelengths below
268 nm, irradiance and lifetime decrease significantly, leading to longer treatment
times. Fig. 1.13 shows the trade-off between germicidal efficacy (i.e., logarithmic
count reduction, LCR), incident irradiance E0, and wavelength using *E. coli*
O157:H7 as an indicator organism for comparison and highlights a currently major
limitation of UV LED technology.

As UVC LED efficiency improves at lower cost of their production and lifetime
increases, they will begin to replace the existing UVC light lamps market as a cost-
effective, environmentally friendly alternative to save energy and improve safety
and shelf life of produced foods from field to fork. It is very likely that in the
near future, many applications that today use mercury lamps will be carried out
by UV LEDs. The wavelength of the commercial UVC LED is in the germicidal
range 240–405 nm, which can enable new applications in existing markets as
well as open new areas. One area that requires more research is their use for odor
removal and volatile organic compound reduction. Success in this area advances
plant safety and helps facilities meet regulatory requirements. In addition, lowering
ethylene levels enhances shelf life.

UV LEDs are more environmentally friendly because they do not generate ozone
and contain no mercury as arc lamps do. They are a cool source compared with arc
lamps, largely because of no output in the infrared range. The ability to instantly turn
the unit off and on enables saving about 50%–75% on electricity.

Conclusions

Germicidal light holds considerable promise for advancing food safety. UV LEDs-based light fixtures have potential to become the driving force behind wider adoption. Because of small chip size, UV LEDs can be installed in areas where installation of current UV lamp equipment is problematic such as cold storage, product transport, and small, countertop, point of preparation, disinfection apparatuses.

However, the fundamental challenge with UVC LEDs is that they are still weak germicidal light generators, and because of their low optical output, they either have to be placed close to where the pathogens reside or run for longer exposure times to produce the requisite pathogen inactivation. Both factors limit their coverage and utility.

To displace the incumbent mercury-based technology, UV LEDs require further development to improve device efficiency and lifetime. For LEDs emitting in the UVC range shorter than 360 nm, the highest efficiency so far has been in the range of a few percent. Also, currently, UVC LEDs are too expensive to qualify as an alternative to traditional UV sources. Economic viability for point of use applications might be achieved within the next years given the technological advances in regard to power output, bulb efficiency, and lifetime predicted by the manufacturers.

UV LEDs application market demands the following improvements from LED manufacturers:

- Higher UV output
- Higher operating efficiency (more electrical input converted to light)
- Lower cost for LEDs and LED system designs more suitable for putting the right amount of light (or radiation) where needed

These market demands are driving rapid technical changes in LED designs; improvement in performance; and reductions in cost.

As device efficiency improves and cost reduces, LEDs will begin to replace the existing UV light sources market and the list of UV LEDs application will certainly grow. The unique properties and performance of UVC LEDs will offer numerous novel and unique approaches to improve safety of food production. Also, UVC LEDs will emerge as environmentally friendly solution to save energy, lower reliance on toxic chemicals, improve worker and consumer safety, extend foods shelf life from field to fork, and improve processing.

References

Andersen, B. M., Bånrud, H., Bøe, E., Bjordal, O., & Drangsholt, F. (2006). Comparison of UV C light and chemicals for disinfection of surfaces in hospital isolation units. *Infection Control; Hospital Epidemiology, 27*(7), 729–734. https://doi.org/10.1086/503643.

Downes, A., & Blunt, T. P. (1878). IV. On the influence of light upon protoplasm. *Proceedings of the Royal Society of London, 28*, 199–212.

Forte, V. C. (2014). *Understanding ultraviolet LED applications and precautions*. Electronic Component News Magazine.

Goodsell, D. S. (2001). The molecular perspective: Ultraviolet light and pyrimidine dimers. *The Oncologist, 6*(3), 298–299. https://doi.org/10.1634/theoncologist.6-3-298.

Green, A., et al. (2018). Inactivation of *Escherichia coli*, Listeria and Salmonella by single and multiple wavelength ultraviolet-light emitting diodes. *Innovative Food Science & Emerging Technologies, 47*, 353–361.

Hijnen, W. A. M., Beerendonk, E. F., & Medema, G. J. (2006). Inactivation credit of UV radiation for viruses, bacteria and protozoan (oo)cysts in water: A review. *Water Research, 40*(1), 3–22. https://doi.org/10.1016/j.watres.2005.10.030.

Jaspers, N. G. J. (2001). Pyrimidine dimers A2. In J. H. Miller, & S. Brenner (Eds.), *Encyclopedia of genetics* (p. 1586). New York: Academic Press.

Kheyrandish, A., Mohseni, M., & Taghipour, F. (2018). Protocol for determining ultraviolet light emitting diode (UV-led) fluence for microbial inactivation studies. *Environmental Science & Technology, 52*(13), 7390–7398.

Koutchma, T., Forney, L. J., & Moraru, C. I. (2009). *Ultraviolet light in food technology: Principles and applications*. CRC Press.

Koutchma, T. (2014). Preservation and shelf-life extension UV applications for fluid foods. 1st edition. Elsevier: Academic Press.

Kowalski, W. (2009). UV Surface Disinfection *Ultraviolet Germicidal Irradiation Handbook: UVGI for Air and Surface Disinfection* (pp. 233–254). Heidelberg: Springer Berlin Heidelberg: Berlin.

Li, J., Hirota, K., Yumoto, H., Matsuo, T., Miyake, Y., & Ichikawa, T. (2010). Enhanced germicidal effects of pulsed UV-LED irradiation on biofilms. *Journal of Applied Microbiology, 109*(6), 2183–2190. https://doi.org/10.1111/j.1365-2672.2010.04850.x.

Li, G. Q., Wang, W. L., Huo, Z. Y., Lu, Y., & Hu, H. Y. (2017). Comparison of UV-LED and low pressure UV for water disinfection: Photoreactivation and dark repair of *Escherichia coli*. *Water Research, 126*(Suppl. C), 134–143. https://doi.org/10.1016/j.watres.2017.09.030.

Mori, M., Hamamoto, A., Nakano, M., Akutagawa, M., Takahashi, A., Ikehara, T., et al. (2007). Effects of ultraviolet LED on bacteria. In *Paper presented at the 10th world congress on medical Physics and biomedical engineering, WC 2006*.

Nakahashi, M., Mawatari, K., Hirata, A., Maetani, M., Shimohata, T., Uebanso, T., & Takahashi, A. (2014). Simultaneous irradiation with different wavelengths of ultraviolet light has synergistic bactericidal effect on Vibrio parahaemolyticus. *Photochemistry and Photobiology, 90*(6), 1397–1403. https://doi.org/10.1111/php.12309.

Physics and Radio-Electronics. (2015). *Light emitting diode (LED). Electronics devices and circuits*. Retrieved August 14, 2018.

Rutala, W. A., Gergen, M. F., & Weber, D. J. (2010). Room decontamination with UV radiation. *Infection Control & Hospital Epidemiology, 31*(10), 1025–1029. https://doi.org/10.1086/656244.

Schubert, E. F. (2006). *Light-emitting diodes*. Cambridge University Press.

SETi. (2012). *Breaks barriers with UVC LED efficiencies of over 10% [press release]. Compound semiconductor*.

Shin, J.-Y., Kim, S.-J., Kim, D.-K., & Kang, D.-H. (2015). Evaluation of the fundamental characteristics of deep UV-LEDs and application to control foodborne pathogens. *Applied and Environmental Microbiology, 59*. https://doi.org/10.1128/aem.01186-15.

Sinton, L. W., Hall, C. H., Lynch, P. A., & Davies-Colley, R. J. (2002). Sunlight inactivation of fecal indicator bacteria and bacteriophages from waste stabilization pond effluent in fresh and saline waters. *Applied and Environmental Microbiology, 68*(3), 1122−1131. https://doi.org/10.1128/aem.68.3.1122-1131.2002.

The Mercury Vapor Lamp. (2016). Retrieved January 3, 2018, from http://lamptech.co.uk/Documents/M1%20Introduction.htm.

The Nobel Foundation. (1903). *The Nobel prize in physiology or medicine 1903. Nobel prizes and laureates*. Retrieved June 14, 2018, from https://www.nobelprize.org/nobel_prizes/medicine/laureates/1903/.

U.S. Army Public Health Command. (2014). *Ultraviolet light disinfection in the use of individual water purification devices*. Army Public Health Center.

U. S. Food and Drug Administration. (2001). 21 CFR Part 179. Irradiation in the production, processing and handling of food. *Federal Register, 65*, 71056−71058.

World Health Organization. (1994). *Environmental health criteria 160: Ultraviolet radiation*. Retrieved Dec. 13, 2017, 2017, from http://www.inchem.org/documents/ehc/ehc/ehc160.htm.

Technology of LED Light Sources and Systems From Visible to UV Range

2

Tatiana Koutchma, PhD

Research Scientist, Guelph Research and Development Center, Agriculture and Agri-Food Canada,
Guelph, ON, Canada

Chapter outline

Abstract

The basic of light-emitting diodes (LEDs) technology and manufacturing features will be presented in this chapter. The fundamental principles of light emission by semiconducting materials, design of LEDs chips and types of packages, requirements to LEDs system are discussed for better understanding of this new light technology. The differences and application areas in production and postharvest processing of fruit and vegetables of visible and UV LEDs are also included to better understand LEDs potential in food industry.

Keywords: LED chip; LED package and LED system; Visible and UV LEDs.

Technology of LEDs

The UV light-emitting diode (LED) technology has been attracting significant attention as a new UV source that can replace conventional mercury gas-filled lamps in curing, printing, and recently disinfection applications. Despite the growth of UV low-pressure lamp food applications, the LED is a relatively new addition to the food production and food treatment toolbox. The manufacturing technology and

Ultraviolet LED Technology for Food Applications. https://doi.org/10.1016/B978-0-12-817794-5.00002-9

mechanism of light generation by LEDs differs from conventional UV lamps. Successful implementation and commercialization of UV LED technology includes better understanding of basic elements of LED technology such as chips and their fabrication, arrays and components of LEDs packages, and system design. This chapter will discuss the essential factors that impact LED performance and system efficiency.

LED Chip

LEDs are solid-state semiconductor devices that convert electrical energy into visible, infrared (IR), or ultraviolet (UV) light depending on the semiconductor material. The energy conversion takes place in two stages: first, the energy carriers in the semiconductor are excited to higher states by electrical energy, and second, most of these carriers, after having lived a lifetime in the higher state, give up their energy as spontaneous emission of photons with energy nearly equal to the bandgap of the semiconductor.

For LEDs to emit light photons, certain elements are combined in specific configurations and electrical current is passed through them. The heart of LEDs is called a "die" or "chip," which is composed of two semiconductor layers—an n-type layer that provides electrons and a p-type layer that provides holes for the electrons to fall into. The actual junction of the layers (called the p-n junction) is where electrons and holes are injected into an active region as shown in Fig. 2.1. There are 12 elements that are important for the construction of the LEDs. B, Si, Ge, As, and Sb serve as base elements; Al, Ga, In, and Sn as p-type dopants; and Ni, P, and C as n-type dopants.

When a voltage source is connected to the LED, electric current flows from the p-side to the n-side (anode to cathode). As the electrons cross the depletion zone and fill a hole, they drop into a state of lower energy. The excess energy is released in the form of a photon that can transport electromagnetic radiation of all wavelengths. The photons are emitted in a narrow spectrum around the energy bandgap of the semiconductor material, corresponding to infrared, visible, and near-UV and UV wavelengths (IESNA, 2005). The color of the emitted light depends on the bandgap energy of the material of the semiconductor. Typically, gallium arsenide is used

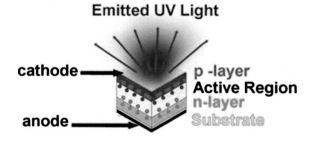

FIG. 2.1

Principle of light emitting by LED chip.

Credit to Phoseon.

for red and IR light; indium gallium aluminum phosphide for green, yellow, orange, and red lights; and gallium nitride and silicon carbide for blue lights. UV LEDs typically are composed of aluminum gallium nitride or indium gallium nitride.

Because of the nature of the gas discharge in a glass envelope, mercury vapor lamps are omnidirectional; that is, the entire surface area of the glass tube can be considered the emitting area. LEDs emit light in a radial pattern from the top of the LED package; that is, the surface area of the LED package can be considered the emitting area.

Similar to blue LEDs, UVC LEDs are grown on 50-mm-diameter wafers, which can hold as many as 1000 individual devices. These wafers are grown epitaxially starting with the n-type layer and finishing with the p-type layer. The method of metal organic chemical vapor phase deposition (MOCVD) is used for growth. The MOCVD technique enables very thin layers of atoms to be deposited on a semiconductor wafer and is a key process or manufacturing III−V compound semiconductors. The wafers are processed in a cleanroom into individual die, singulated, and flipped onto a submount where electrical connections are bonded to the die.

Temperature and current are two major factors that affect the lifetime of LEDs. As LEDs convert electricity into light, heat is also created within the p-n junction, known as the junction temperature. For a diode to achieve maximum life expectancy, the junction temperature has to remain in a safe operating zone. Heat must be effectively conducted away from the die by the packaging materials or the device leads. Without proper thermal management, internally generated heat can cause packaged LEDs to fail. The UV power output of a diode increases with input current but decreases with junction temperature. At any fixed input current, the cooler the junction temperature remains, the more UV output power the diode would provide.

LED's chip sizes can range from tenths of millimeters for small-signal devices to greater than a square millimeter for the power packages available today. Currently, the largest-power LEDs on the market have emitting area of 12 mm^2 with rectangular chip in the center of the package. The large copper submount is used for adequate thermal management and is also equipped with a thermistor for thermal sensing (RadTech, 2016). The specific design of an individual UV LED chip or diode depends on the desired wavelength, peak UV irradiance, and capabilities of the LED chip manufacturer.

Early successes were achieved with longer wavelength in visible light and infrared. UV LED with an efficiency of around 1% was first produced in a lab environment in Japan in 1992. The technical challenges for the design and development of UV LEDs range from learning how to prepare high crystalline quality and low defect density epitaxial layers of the required compounds on top of CIS AlN substrates, achieving low resistance, good electric contacts, and packaging the device.

Packaged LEDs

Because of the complexity of wafer and chip fabrication processes, the light output power of UV or visible LED can vary by a factor of 2, even for chips coming from the

Table 2.1 Types of LED Packages.

LED Type	T-Pack	Surface Mount	Chip on Board
Device Image			
Packed Array (10 × 10 mm)			
Density	9 LEDs	40 LEDs	342 LEDs
Array Power	0.4 W	4 W	68 W

Used with permission from AquiSense Technologies, KY, USA.

same wafer. Binned chips are sold to a packager where the chip is placed in a protective package with optics and solderable leads. By using multiple chips or groups of packaged LEDs, the intensity and power of light needed by the application can be obtained. The typical packaged LED system contains multiple elements stacked together for optimum performance. As shown in Table 2.1, there are three types of LED packages:

- Through hole or T-Pack style
- Surface mount device (SMD)
- Chip on board (COB)

The majority of UV LEDs are now surface mount diodes or SMDs. The SMD may be a single or multiple (usually no more than 4) chips inside a ceramic package, which can be soldered onto a printed circuit board. Chip on board (COB) LEDs may be mounted directly onto the circuit board but require more specialized equipment and are usually only used when high packing density is needed to achieve higher intensity or the working distance is small compared with the area to be irradiated.

The examples of high-power blue LED transistor T-pack and SMD chips from different manufacturers are shown in Fig. 2.2.

A basic LED package can consist of one or several chips, while an LED array usually contains a large number of chips arranged in a matrix pattern. Both terms refer to an arrangement of LEDs that are physically and electrically assembled together and include a means for the entire assembly to be electrically connected to another device. When a heat sink and, occasionally, optics are added to the LED package, it is often referred to as an LED module. Several LED modules can be joined together to form an even larger LED array. As shown in Fig. 2.3, the power density of an LED array is a function of LED chip optical power and packing density of the array.

Driver circuitry, thermal management, mechanical mounting, controls, lenses, and optics are presented as an optimized LED system package.

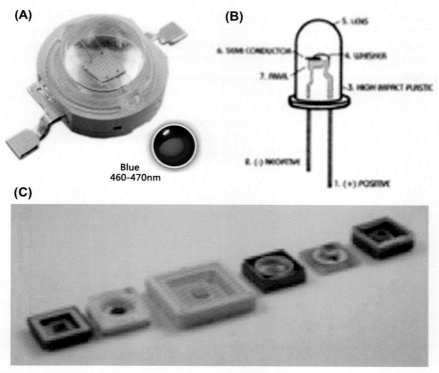

FIG. 2.2

High-power discrete blue LED (A) and cross section (B), SMD chips from various manufacturers (C).

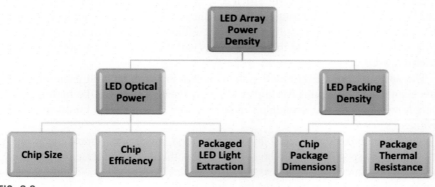

FIG. 2.3

Elements defining LED array power density.

1. LED driver or current regulator converts the supply voltage to a DC voltage and provides DC power supply and control functions. It holds the current at a constant level/output over variable supply voltage ranges. A single DC power source may drive one LED or a cluster of LEDs.
2. Heat sink provides thermal management to optimize system performance.
3. LED arrays: high power LEDs are assembled onto the metal core circuit board to provide mechanical, electrical, and thermal connections.
4. Collimators direct lights and produce various beams width.
5. Optics create additional beams and various distribution patterns. Typically, an LED package uses an optically clear material (encapsulant) to form a lens atop the LED chip.

The packaged LEDs are supplied to the system integrators that add electronics, thermal management, optics, and housings to create a finished module.

LED Systems

LED manufacturers typically integrate either bare chips (B) or packaged chips (C) to build an LED system. The targeted application will define the wavelengths of LEDs, the UV dose required, and UV irradiance. The wavelength selection is essential to getting the most from UV LED disinfection systems. Phoseon discovered utilizing both 275 and 365 nm wavelengths provides a synergistic effect allowing even faster, stable reactions.

Another biggest challenge for system integrators is that although LEDs can be very efficient, a single, packaged LED may not produce much optical output in the visible or UV range. To get higher irradiance for specific UV applications, many LED packages are typically combined into a single fixture.

Fig. 2.4 illustrates concept of disinfection of the conveyor belt using UV LED technology to inactivate pathogens after the food has passed through the conveyor.

To achieve maximum system efficiency, the integration of LEDs into UV systems has to be done considering the performance of UV source including individual LED chips (wavelength, cost, power output, efficiency, and lifetime); LED packaging and encapsulation; and thermal controls with the performance of the chamber (reactors) itself.

The advantage of LEDs-based systems is that they can be built using multiple-wavelengths LED modules for higher efficacy. Each module can be outfitted with a single wavelength and assembled in such a way as to provide multiwavelength coverage.

Ideally, engineers and LED buyers and food processors should be directly involved in the specification of the LEDs, drivers, thermal management technology, and other system components to assure that the end LED product will have the performance and lifetime required for the applications.

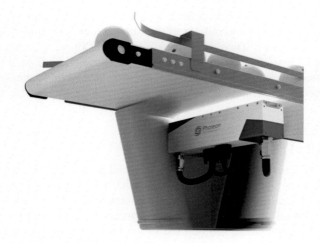

FIG. 2.4

UV LEDs disinfection of food conveyor.

Credit to Phoseon.

LED Light Sources and Food Applications

LED sources can provide narrow discrete emission range in the full spectrum of light radiation: UVA (315–400 nm), UVB (280–315 nm), and UVC (200–280 nm), visible light (400–760 nm), and infrared light (760–3000 nm).

Visible LEDs

In the case of visible light, the applications of LEDs are mainstream and ubiquitous. They are present in practically every home appliance, traffic lights, publicity signs, flashlights and even car lights, etc. LEDs are cheap, lightweight, small, and rugged. Red LEDs show efficiencies in the order of 80% with lifetimes exceeding 100,000 h.

Also, visible LEDs present a solution for the natural crop cultivation in hydroponic farming and greenhouse crop production. Their benefits are reported in a substantial number of studies that have demonstrated the usefulness of LEDs in the crop cultivation (D'Souza, Yuk, Khoo, & Zhou, 2015). The benefits of LED that are valuable in horticultural production include the ability to control and fine-tune the quality of light, to limit the amount of heat generated, as well as to ease integration into electronic systems to give greater control over the emitted light. It has been shown that light stimulates the production of various nutrients, antioxidants, and secondary metabolites in plants, which function to provide defense against reactive oxidation species (ROS) produced during photosynthesis or light stress. Another important area of applications of visible and blue light LEDs is the control of the rate of ripening during transportation and storage. Depending on the wavelength (red, blue, or white), light has varying effects on different types of fruits and vegetables and delays or accelerates ripening and result in better appearance or texture attributes.

Infrared (IR) LEDs

An IR LED, also known as IR transmitter, is a special purpose LED that transmits IR rays in the range of 760 nm wavelength up to 3000 nm. Such LEDs are usually made of gallium arsenide or aluminum gallium arsenide. They, along with IR receivers, are commonly used as sensors. The appearance of IR LEDs is similar to a common LED. The camera can be used as the IR rays being emanated from the IR LED in a circuit.

UV LEDs

Despite the advances in visible and IR range, further expansion of LEDs in UV spectrum still presents many challenges. Unlike visible light LEDs, UV LEDs that have wavelengths less than 350 nm are not grown on gallium nitride (GaN) substrates but rather are produced on aluminum nitride (AlN) substrates. Also, UVB and UVC LEDs require specialized tools for production, which greatly raises barriers for entry when compared with their UVA counterpart.

Also, UV LED chips are currently less efficient than conventional UV lamps systems as well as visible and IR LEDs. Visible LEDs, which require little or no cooling, claim to last indefinitely and are 80%–90% more efficient than normal lights and are relatively inexpensive. Existing LED technology limitations render UV LEDs around 10%–20% efficient for longer wavelengths of 395 and 405 nm and less than 10% for shorter wavelengths less than 365 nm. This occurs due to the fact that UV LED chips have not yet been optimized for the UV region and large amount of energy is lost in the form of heat. The amount of heat energy is so significant that the only way to effectively remove it from the system is by circulating a liquid coolant around a heat sink attached to the chips. Air is also used for cooling purposes. Decrease in relative UV intensity peaks in shorter wavelength region shown in Fig. 2.5 demonstrates this trend. The large drop in LED efficiency at wavelengths below 375 nm (Fig. 2.4) will require significant advancement in LED chip technology.

With continued development in the material science and manufacturing process, UV LEDs should increase in optical output, electrical efficiencies, and lifetime with decreasing cost of production.

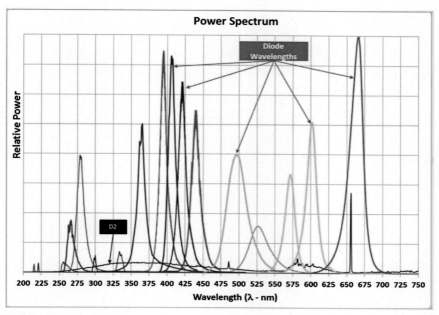

FIG. 2.5

LEDs spectral distribution versus relative intensity in UV, visible, and infrared range.

Credit to Phoseon.

References

D'Souza, C., Yuk, H. G., Khoo, G. H., & Zhou, W. (2015). Application of light-emitting diodes in food production, postharvest preservation, and microbiological food safety. *Comprehensive Reviews in Food Science and Food Safety, 14*, 719–740.

IESNA. (2005). *Technical memorandum on light emitting diode (LED) sources and system.* TM-16-05-2005.

RadTech International. (2016). *UV LED curing technology, e-book.* Radtech.org.

Applications and Advances in LEDs for Horticulture and Crop Production

Akvilė Viršilė[1], Giedrė Samuolienė[1], Jurga Miliauskienė[1], Pavelas Duchovskis[2]

Doctor, Lithuanian Research Centre for Agriculture and Forestry, LT-54333, Babtai, Lithuania[1];
Professor, Lithuanian Research Centre for Agriculture and Forestry, LT-54333, Babtai, Lithuania[2]

Chapter outline

Abstract

The effects of UVA and UVB light on leafy vegetables, herbs, and fruiting greenhouse crops are explored as well as UV perception and response mechanisms are reviewed. The emerging possibilities to apply specific wavelengths of UV LEDs for plant lighting enabled for targeted research and applications. Properly balanced UV LED parameters (wavelength, intensity, timing, and duration of exposure), consistent with other environmental factors, might create plant species-specific eustress conditions and therefore can be used to improve plant growth, morphology, and phytochemical quality.

Keywords: Green vegetables; Growth; Herbs; Light-emitting diodes; Phytochemicals; Seedlings; UVA; UVB.

Introduction

In recent years, consumption of fresh vegetables and herbs has increased in daily diets. The growing demand for high-quality vegetables is forcing to encounter to controlled environment agriculture as an alternative and supplement to field production (Dou, Niu, Gu, & Masabni, 2017). In field production, varying climatic conditions among seasons or locations and changing environmental conditions lead to varying vegetable yield and nutritional value (Zobayed, Afreen, & Kozai, 2005). As an alternative, growers turn to controlled environment agriculture including tunnels, greenhouses, and indoor vertical farms. The controlled environment technologies allow the precise control of optimized environmental conditions ensuring stable, maximal, and high-quality biomass and phytochemical production, free from biotic and abiotic contamination (Castilla & Herandez, 2006; Despommier, 2010; Dou et al., 2017; Kozai, 2013; Kozai, Niu, & Takagaki, 2015; Zobayed et al., 2005).

Scientific researchers indicate that light quality is one of the most important environmental factors affecting vegetable growth and development, also influencing their internal quality (Amaki, Yamazaki, Ichimura, & Watanabe, 2011; Dou, Niu, Gu, & Masabni, 2017; Samuolienė et al., 2011; Shiga et al., 2009). Controlled environment systems usually use artificial lights to ensure plant growth and high quality. New lighting technologies such as light-emitting diodes (LEDs) have the capacity to meet the light intensity, dose, and wavelength requirements of different plant species (Amaki et al., 2011; Brazaitytė et al. 2016; Bugbee, 2016).

LED technology for horticultural applications made significant progress in recent years and is continuously developed. It enabled to control the light spectrum, thus creating different lighting conditions optimal for various plants. This provided new opportunities for targeted regulation of plant growth and metabolic responses seeking to optimize productivity and quality under controlled environments (Bantis et al., 2018; Carvalho, Schwieterman, Abrahan, Colquhoun, & Folta, 2016; Dou, 2017).

The concept of the prime lighting spectrum recipe has significantly changed over the years. In the first publications on LED lighting for plants, the combinations of red and blue light were analyzed as the most efficient (Massa, Kim, Wheeler, & Mitchell, 2008; Olle & Viršile, 2013). Following the McCree experiments, red and blue light was proved to be absorbed by leaves better than other regions of visible light spectrum (Bugbee, 2016). Herewith, red and blue LEDs were and still are characterized by relatively highest photon efficacy (Nelson & Bugbee, 2014). Therefore, the first industrial horticultural LED light sources were composed of red and blue diodes. However, during the years of investigations, the remarkable effects of supplemental light colors, as green, far red, and UVA (Mitchell et al., 2015), were explored. Lighting strategies changed from solid red and blue combinations to multicomponent lighting spectrum, mimicking wide spectrum of the sun. In recent years, the advance in UV LED technology opens the door for the new level of photobiological research, enabling to use UV light as the eustressor in horticultural crop production

(Bassman, 2004; Neugart & Schreiner, 2018; Wargent, Moore, Roland Ennos, & Paul, 2009).

Natural solar ultraviolet (UV) radiation (280−400 nm) was frequently analyzed as injurious source of stress for plants. High dose UVB radiation (280−320 nm) results in altered expression of genes involved in biosynthesis of phenols, as well as in the massive production of reactive oxygen species (ROS). ROS production in a short time may override the antioxidant capacity, leading to a severe UVB distress, which can induce programmed cell death (Hideg, Jansen, & Strid, 2013; Jordan, 2002; Müller-Xing, Xing, & Goodrich, 2014). Low dose UVB radiation can induce alterations in antioxidant status, for example, regulation of glutathione pathways, phenylpropanoids, cinnamates, or flavonoid pathways, and pyridoxine biosynthesis pathways. The data of current research offer possibilities to use UV light as regulatory signal (Wargent, 2016), affecting development and metabolism (Czégény, Mátai, & Hideg, 2016).

Adapting the possibilities of artificial lighting, relatively small amounts of UV might be used in horticulture for different purposes. Notwithstanding that UV is outside of the photosynthetically active waveband, it is still biologically active and regulates photomorphogenetic (Jenkins, 2017), morphological (de Carbonnel et al., 2010; Gardner, Lin, Tobin, Loehrer, & Brinkman, 2009; Wargent et al., 2009) as well as metabolic (Agati et al., 2013; Mewis et al., 2012; Neugart et al., 2012; Park et al., 2007; Schreiner et al., 2012; Tossi, Amenta, Lamattina, & Cassia, 2011) plant responses.

Current research knowledge on UV light effects on plants is substantiated on various broad emission spectrum light sources or light filters. Most popular horticultural light sources have relatively low flux of UV radiation: from 0.01% of photosynthetic photon flux (PPF) of UVB and 0.8% PPF of UVA in high-pressure sodium lamps to 0.13% PPF of UVB/8.2% PPF of UVA in metal halide and 0.49% PPF of UVB and 2.1% of UVA in very high output fluorescents lamps. For comparison, sunlight on the noon of May, on clear day, has 0.47% and 8.5% of PPF of UVB and UVA, respectively (Nelson & Bugbee, 2013). The LEDs emitting in UVA and UVB range also appeared in the market, and initial research has been conducted seeking to explore the effects of specific UV wavelengths on plants (Kneissl & Rass, 2016; Wargent, 2016). Compared with UVB, the number of studies on the effects of UVA is limited mainly due to the experimental design that often does not allow studying the effects of UVA independently from the UVB. Recently, LEDs offer the possibility to tailor the light spectrum, which will enable to explore the effects of individual wavelengths of UVA, UVB, and blue light (Neugart & Schreiner, 2018). As summarized in Table 3.1, the longer wavelengths of UVA light (400−380 nm) are applied in the research improving growth and metabolism of horticultural plants. The intensity of UVA lighting varies from 1% to 4.2% and 10.4% of total PPF for basil (Bantis, Ouzounis, & Radoglou, 2016; Brazaitytė et al., 2016; Vaštakaitė et al., 2015), about 4.5% of total PPF for tomato seedlings (Brazaitytė et al., 2010), and 6% for lettuce (Li & Kubota, 2009) with the duration matching with photoperiod of general lighting. Albeit, low photon efficacy and fast aging of

Table 3.1 Effects of UV LEDs on Plants Metabolism and Growth Characteristics.

Properties of Supplemental UV	Background Lighting Conditions	Plant Species	Effects on Plants	References
UVA LEDs (400 nm)	Kale cultivated under cool-white fluorescent lamps, PFD 275 μmol m^2 s^{-1}, 16 h for 7 days; after that supplemented with 730, 640, 525, 440, or 400 nm, PFD 15.2, 253.3, 6.5, 10.6, and 6.9 μmol m^{-2} s^{-1}, respectively.	Kale (*Brassica oleracea*. L. cv. Winterbor)	Contents of lutein, β-carotene, chlorophylls *a* and *b* were not statistically different or decreased.	Lefsrud et al., 2008
UVA LEDs at <400 nm	20% blue light at 400–500 nm + 39% green light at 500–600 nm + 35% red light at 600–700 nm + 5% far-red light at 700–800 nm (red: far-red—8.16) + 1% UV light at <400 nm. PFD—200 ± 20 μmol m^{-2} s^{-1}.	Basil (*Ocimum basilicum* L. cv. Lettuce Leaf and Red Rubin-mountain Athos hybrid	Supplemental 1% UV increased total biomass, root:shoot ratio, total phenolic content.	Bantis et al., 2016
UVA LEDs (402, 390, and 366 nm)	Blue (447 nm), red (638 nm), deep red (665 nm), far red (731 nm) supplemented with UV LEDs emitting at 366, 390, 402 nm. PFD of blue, red, deep red, and far-red LEDs—300 μmol m^{-2} s^{-1} with different regimes in the UV LEDs PFD—6.2 μmol m^{-2} s^{-1}, 12.4 μmol m^{-2} s^{-1}.	Basil (*Ocimum basilicum* L. cv. Sweet Genovese)	Supplemental UVA LEDs at higher intensity level increased leaf area (366 and 402 nm) and height (402 nm). Supplemental UVA LEDs at lower intensity increased fresh weight (366 and 402 nm), leaf area (366 nm), height (402 nm), and hypocotyls length (390 and 402 nm). All supplemental UVA slightly or significantly increased DPPH (2,2-diphenyl-1-picrylhydrazyl) free-radical scavenging activity. Supplemental UVA at higher intensity level increased total phenol content (366, 390, and 402 nm), α-tocopherols increased (366 nm) or decreased (390, 402 nm), total anthocyanins (390 and 402 nm) and ascorbic acid (366 nm) decreased. Supplemental UVA LEDs at lower intensity decreased α-tocopherols (366, 390, and 402 nm) and ascorbic (366 nm) acid content.	Brazaitytė et al., 2015

UVA LEDs	Light treatment	Plant species	Effect	Reference
UVA LEDs (395 nm)	80% red at 657 nm + 20% blue at 447 nm PAR + UVA at 395 nm.	Basil (Ocimum basilicum L.)	Addition of UVA to red and blue light had significant effect on the chilling tolerance and shelf life performance.	Jensen et al., 2018
UVA LEDs (390 nm)	HPS lamps and natural daylight supplemented with UVA (390 nm, ~13.0 μmol m^{-2} s^{-1}); total PFD ~125 μmol m^{-2} s^{-1}.	Basil (Ocimum basilicum L. purple-leaf cv. Dark Opal and green-leaf cv. Sweet Genovese)	UVA improved antioxidant properties, increased total phenol and anthocyanin contents, ascorbic acid concentration, ABTS radical scavenging activity. More expressed effect on green-leaf basils.	Vaštakaitė et al., 2015
UVA LEDs (385 nm)	Lentil seeds were germinated under blue (455 nm), red (638, 669 nm), far-red (731 nm), PFD 100 μmol m^{-2} s^{-1}; 12 h; or B + R + FR supplemented with 4 μmol m^{-2} s^{-1} UVA LEDs.	Spouted lentil seeds (Lens esculenta Moenh.)	Content of total phenols increased, whereas vitamin C and α-tocopherol decreased under supplemental UVA LEDs.	Samuolienė et al., 2011
UVA LEDs (380 nm)	Closed LED boxes with emission wavelengths of 380 (UVA), 450 (blue), 470 (blue), and 660 (red) nm.	Ginseng roots (Panax ginseng Meyer)	UVA increased the concentration of protopanaxadiol ginsenosides Rb$_2$ and Rc. The ratio of protopanaxadiol-type ginsenosides to protopanaxatriol-type ginsenosides was changed.	Park et al., 2012
UVA LEDs (380 nm)	Combination of 447, 638, 669, and 731 nm LEDs, supplemented with 9 μmol m^{-2} s^{-1} of 380 nm UVA (total PFD 200 μmol m^{-2} s^{-1}), 16 h.	Tomato (Lycopersicon esculentum L., Raissa F1 seedlings)	Supplemental UVB enhanced tomato growth: total fresh weight, hypocotyl diameter, leaf area were increased.	Brazaitytė et al., 2010
UVA LEDs (380 nm)	Combination of 447, 638, 669, and 731 nm LEDs, supplemented with 9 μmol m^{-2} s^{-1} of 380 nm UVA (total PFD 200 μmol m^{-2} s^{-1}), 18 h.	Cucumber (Cucumis sativus L. cv. Mandy F1 seedlings)	Supplemental UVA LED light decreased growth and development of cucumber seedlings.	Brazaitytė, et al., 2009
UVA LEDs (376 nm)	658 nm red LEDs (PFD 90 μmol m^{-2} s^{-1}), supplemented with two levels of UVA light: red + low UV (3.2 W·m^{-2}) and red + high UV (6.8 W·m^{-2}).	Tomato (Lycopersicon esculentum L. cv. Superdoterang seedlings)	Under both UVA treatment levels, tomato seedlings were more compact, the growth of plant organs was balanced, leaf area was increased, and total plant fresh and dry weights were also enhanced.	Khoshimkhujaev et al., 2014

Continued

Table 3.1 Effects of UV LEDs on Plants Metabolism and Growth Characteristics.—*cont'd*

Properties of Supplemental UV	Background Lighting Conditions	Plant Species	Effects on Plants	References
UVA LEDs (373 nm)	Cool white fluorescent lamps, PFD 300 μmol m^{-2} s^{-1}, 16 h, supplemented with 18 μmol m^{-2} s^{-1} UVA LEDs.	Red leaf lettuce (*Lactuca sativa* L. cv. Red Cross)	Supplemental UV resulted in increased anthocyanins content, but decreased carotenoids, phenolic compounds, and growth characteristics.	Li & Kubota, 2009
UV-A LEDs (365 nm)	Broccoli, cultivated under fluorescent lamps, PFD 100 μmol m^2 s^{-1}, 16 h for 4 weeks; supplemented with 50 μmol m^{-2} s^{-1} each for 515, 470, 420, and 61 μmol m^{-2} s^{-1} for 365 nm.	Broccoli (*Brassica oleracea* L. cv. Monopoly)	Concentration of 3-indolylmethyl glucosinolate was increased under UVA treatment, but contents of flavonol glycosides decreased.	Rechner et al., 2017
UVA LEDs (340 and 325 nm)	Lettuce, cultivated under white fluorescent lamp, PFD 150 μmol m^{-2} s^{-1}, 16 h, 3 days before harvest was transferred under red LEDs 660 nm, supplemented with 325 or 340 nm UV (0.5 W m^{-2} for 16 h).	Red leaf lettuce (*Lactuca sativa* L. cv. Red Fire)	Increased total ORAC values under supplemental 325 > 340 nm LEDs, compared with white light, but the effect was lower compared with supplemental UVB 310 nm LEDs.	Goto et al., 2016
UVB LEDs (310 nm)	Lettuce, cultivated under white fluorescent lamp, PFD 150 μmol m^{-2} s^{-1}, 16 h, 3 days before harvest was transferred under red LEDs 660 nm, supplemented with 310 nm UV (0.5 W m^{-2} for 16 h).	Red leaf lettuce (*Lactuca sativa* L. cv. Red Fire)	Anthocyanins, 3-O-glucosyltransferase expression increased compared with white light and to the effect of supplemental 325 and 340 nm.	Goto et al., 2016

LED, *light-emitting diodes*; PAR, *photosynthetically active radiation*; PFD, *photon flux density.*

current UV LED technology still delay their incidence in industrial horticultural applications (Nelson & Bugbee, 2014; D'Souza et al., 2015; Bugbee, 2017).

UV Light Perception

UV radiation is divided into three wavebands: UVA (320–400 nm), UVB (280–320 nm), and UVC (100–280 nm). UVC is fully absorbed by the stratospheric ozone layer and the atmosphere (Huché-Thélier et al., 2016). At the sea level, UV component comprises 6% of the solar radiant energy, and it contains 10–100 times more UVA than UVB (Verdaguer, Jansen, Morales, & Neugart, 2017). The levels of UVA and UVB vary with latitude, season, time of the day, weather conditions, etc. However, the diurnal and seasonal variation of UVB is remarkably higher than that of UVA. Consequently, the UVA/UVB, as well as UVB/PAR ratio, is highly variable (Neugart & Schreiner, 2018).

Plants have particular photoreceptors that perceive UV radiation. UVB radiation is perceived by specific receptor UV RESISTANCE LOCUS 8 (UVR8), and there is no real evidence that UVA can evoke UVR8-mediated responses (Christie et al., 2012; Kataria, Jajoo, & Guruprasad, 2014; Rizzini et al., 2011; Verdaguer et al., 2017; Wu et al., 2012; Yin & Ulm, 2017). UVA radiation, together with blue light, activates phototropins, cryptochromes (peak absorption at 370 and 450 nm), and Zeitlupe (ZTL) proteins (Huché-Thélier et al., 2016). Phytochrome, efficiently activated by red or far red light, also absorbs in wide spectrum of 300–800 nm at lower rate; therefore, it can also be activated by the part of UV spectrum (Huché-Thélier et al., 2016). On the basis of current knowledge, the perception of UVB signal is highly specific, when UVA signaling is tightly connected with blue light signaling. These principles validate the high level of integrity in light signaling in higher plants and also support the diversity of UVB and UVA radiation effects.

Photoresponse to UV Light

Highly energetic shorter wavelengths of UVB radiation can potentially induce several deleterious effects in plants, including disruption of the homeostasis, oxidative damage, changes in plant biochemistry, partial inhibition of photosynthesis, reduction in growth, and detrimental effects on the important macromolecules—DNA, proteins, and lipids (Hollósy, 2002; Kataria et al., 2014; Neugart & Schreiner, 2018; Verdaguer et al., 2017). Plants acclimate to the tolerable dose of UVB radiation seeking to minimize injurious UV effects: they form smaller, but thicker leaves, accumulate lower biomass, are shorter in height, modify photosynthesis, and accumulate less of chlorophylls but more of phenolic compounds, which function as UV protecting components, activating the response of antioxidant protective system (Fig. 3.1) (Agati et al., 2013; Ballare, Caldwell, Flint, Robinson, & Bornman, 2011; Hideg et al., 2013; Jansen & Bornman, 2012; Neugart & Schreiner, 2018;

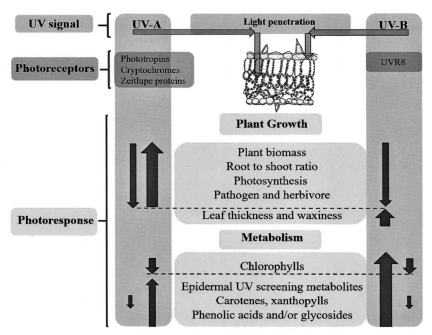

FIGURE 3.1

The effect of UV light on plant growth and metabolism.

Robson, Klem, Urban, & Jansen, 2015; Suchar & Robberecht, 2015). The contents of other secondary plant metabolites, as glucosinolates and carotenoids, were also reported to be affected (Schreiner et al., 2012). These changes in plant properties affect the quality of fruit and vegetable production, including biomass production, external and internal quality, or organoleptic properties (Schreiner, Korn, Stenger, Holzgreve, & Altmann, 2013).

UVA radiation role in plant physiological processes is less explored and currently is still described as the "known unknown" (Verdaguer et al., 2017). The results of research performed are often contradictory. It is reported that UVA has strong morphological effects, especially on the reduction of leaf area. However, the UVA effect on biomass production and net photosynthesis is species specific (Baroniya, Kataria, Pandey, & Guruprasad, 2014; Neugart & Schreiner, 2018; Verdaguer et al., 2017), and information on UVA effects on secondary plant metabolites is limited (Lee, Son, & Oh, 2014; Li & Kubota, 2009). UVA is reported to be less efficient than UVB in mediating specific biological responses; however, as the UV penetration through leaf tissues increases as wavelength increases, UVA can reach much deeper tissues in the leaf and penetrate into the canopy (Verdaguer et al., 2017). Therefore, its regulating properties are apparent both for individual plants and plant communities and are expectant for further investigation.

Plant physiological responses to UVB and UVA radiation are modified by radiation dose, environmental factors (as PAR, background lighting spectrum, humidity, temperature, etc.), as well as depend on plant species and chemotype (quantity and structure of flavonoids, accumulated in plant tissues) (Escobar-Bravo, Klinkhamer, & Leiss, 2017; Hoffmann, Noga, & Hunsche, 2015; Neugart & Schreiner, 2018; Verdaguer et al., 2017). Research results are also often controversial, as the results tightly depend on the experimental design: intensity, duration and timing of the UV treatment, spectrum, and PAR level of the main lighting source. Furthermore, the current research knowledge on the effects of UV LED on plants of different life strategies will be overviewed and the main trends will be discussed on plant physiological basis.

UVA and UVB Effects on Green Leafy Vegetables

Over the past two decades, the demand for the quality of horticultural products is constantly rising: the society prefers the fresh products of high organoleptic, nutritional, and functional quality (Rouphael, Kyriacou, Petropoulos, De Pascale, & Colla, 2018). Green vegetables of stable productivity, fine morphology and coloration, minimized biotic and abiotic contamination, new health benefits, and increased shelf life can be produced in controlled environment systems (Mahajan et al., 2017; Nicole et al., 2016). Supplemental light is a useful modulator of vegetable quality (Rouphael et al., 2018)—light spectrum is an important environmental signal, regulating plant physiological and biochemical processes. The effects of blue, red, and green light have been investigated with various green vegetable species (Bian, Yang, & Liu, 2014; Samuolienė, Brazaitytė, & Vaštakaitė, 2017; Viršilė, Olle, & Duchovskis, 2017). Here, the impacts of UV irradiation and cases of UV LED applications will be explored.

The general tendency of UV radiation on lettuce includes inhibited growth or morphological changes (Paul, Jacobson, Taylor, Wargent, & Moore, 2005; Tsormpatsidis, Henbest, Battey, & Hadley, 2010; Wargent, Elfadly, Moore, & Paul, 2011). Experiments using films varying in their ability to transmit UV radiation demonstrated decrease in growth parameters of various lettuce cultivars (Krizek, Britz, & Mirecki, 1998; Tsormpatsidis et al., 2008). Hectors, van Oevelen, Guisez, Prinsen, and Jansen (2012) reported auxin-dependent flavonoid patterns in response to UVB radiation, thus greatly contributing to UVB-induced morphogenic responses, whole-plant architecture, where more axillary branching was associated with an increase in UVB absorbing compounds under low UVB doses.

Contrary to discussed results, Fina et al. (2017) demonstrated that UVB levels present in solar radiation inhibit maize leaf growth without causing any other visible stress symptoms, including accumulation of DNA damage. Authors state that decrease in leaf growth in UVB-irradiated leaves is a consequence of a reduction in cell production and a shortened growth zone. A number of studies have shown that stomatal opening and closure are highly related to UV radiation and depends on wavelengths: UVA stimulates opening (Jansen & Van Den Noort, 2000), but UVB radiation distinguishes in controversial results (Teramura, Tevini, & Iwanzik, 1983; Tossi et al., 2011).

The general tendency emerging from reported data is that changes in leaf anatomy and decrease in biomass occur due to exceeded limits of plant tolerance to UV dosage (Davis & Burns, 2016; Piri, Babaeian, Tavassoli, & Esmaeilian, 2011).

One of the main current objectives in controlled environment horticulture is to optimize yield and quality of various green vegetables. Currently performed plant photoresponse research has been addressing quality and growth efficiency separately (Nicole et al., 2016). As reviewed further in the text, the UV lighting effects on phytochemical contents and metabolic processes in green vegetables are also widely investigated; however, the effects on photosynthetic aspects are equally important.

Wargent et al. (2011; 2015) demonstrated an increase of the net photosynthesis in lettuce after exposure to high-light UVB, suggesting that photosynthetically active radiation and UVB might have a synergistic and positive effect on plant photoprotection. Tsormpatsidis et al. (2010) also support the opinion that phenolic compounds effectively protect the photosynthetic apparatus because ambient levels of UV radiation did not limit the efficiency of photosynthesis, whereas reduction of photosynthetic rate of many species in closed environment under increased UVB radiation was reviewed by Kakani, Reddy, Zhao, and Sailaja (2003). The positive effect on photosynthetic pigment accumulation of 16 green and red leaf lettuces was obtained under UVB radiation regarding UVA (Caldwell & Britz, 2006). Such response might be explained by higher levels of phenolic compounds, which reduce the photoprotective requirement for chloroplast carotenoids (Britz, Caldwell, Mirecki, Slusser, & Gao, 2005).

As it was mentioned above, the biosynthesis of secondary metabolites is closely related to plant defense mechanisms and is induced in response to stresses, e.g., UV radiation (Agati et al. 2013). One of the biggest groups of bioactive compounds sensitive to UV radiation is thought to be flavonoids, as they serve important roles during the establishment of plants based on UVB screening capacities. But flavonoid biosynthesis and the flavonoid absorption spectra still pose some concerns on their primary functions as UVB screeners (Cockell & Knowland, 1999). It seems that UVB radiation is not a prerequisite for the biosynthesis of flavonoids (Jenkins, 2009). There are literature data, showing that the very same "antioxidant" flavonoids accumulate because of sunlight in the absence or in the presence of UV radiation (Agati et al., 2011; Bilger, Rolland, & Nybakken, 2007). On the other hand, Huché-Thélier et al. (2016) review similarities in the plant responses triggered by UVB or blue light signals and support the hypothesis of synergistic effect between both light components on the production of UVB-induced flavonoids.

Lettuce has been used as a model crop in terms of secondary metabolic responses to UV exposure in many studies. Curvilinear relationship between the anthocyanin content and UV radiation was demonstrated by Tsormpatsidis et al. (2008) using UV selective films. Krizek et al. (1998) found that UVB radiation is more important than UVA radiation for flavonoid induction in this red-pigmented lettuce cultivar.

Ebisawa et al. (2008) demonstrated the increase of quercetin content and flavonol synthase gene expression in lettuce under supplementary UVB from fluorescent lamps together with blue fluorescent lamp light at night. An increase of anthocyanin production in lettuce under combination of UVA or UVB with blue light was also reported by Park et al. (2007). Moreira-Rodríguez, Nair, Benavides, Cisneros-Zevallos, and Jacobo-Velázquez (2017) found that broccoli sprouts exposed to UVA or UVB radiation accumulated more specific glucosinolates, phenolics, carotenoids, and chlorophylls compared with control. Increased concentrations of quercetins, kaempferols, and chlorogenic acid were presented by Tegelberg, Julkunen-Tiitto, and Aphalo (2004) under combined supplemental UVB and far-red light. However, UVB-mediated induction of plant flavonoids can depend on the far-red light doses, as UVB perception by plants blocks the signal transduction triggered by low-red and far-red conditions (Hayes, Velanis, Jenkins, & Franklin, 2014; Mazza & Ballaré, 2015).

Research, performed with individual UV wavelengths, allows indicating more specific UV impacts (Table 3.1). Li and Kubota (2009) demonstrated an increase of anthocyanins with supplemental UVA 373 nm LEDs to cool white fluorescent lamps in baby leaf lettuce cv. Red Cross, but accumulation of carotenoids and phenolic compounds and growth characteristics were suppressed. Supplemental UVA 385 nm LEDs to blue, red, and far- red LED light affected an increase of total phenols in spouted lentil seeds but had no effect in sprouted radish or wheat seeds (Samuolienė et al., 2011). Lefsrud, Kopsell, and Sams (2008) reported that UVA (400 nm) LEDs, supplemental for fluorescent lamps, had no positive effect on accumulation of lutein, β-carotene, chlorophylls *a* or *b* in kale, compared with supplemental far-red, red, green, or blue LED treatments (Lefsrud et al., 2008). Goto, Hayashi, Furuyama, Hikosaka, and Ishigami (2016) presented a significant increase of anthocyanins and 3-O-glucosyltransferase expression in red leaf lettuce cv. Red Fire, treated 3 days before harvest with supplemental UVB LEDs (310 nm) compared with supplemental UVA LEDs (325 and 340 nm) treated plants, whereas total oxygen radical absorbance capacity of all the UV treatments was higher than control. Continuous UVA, repeated or gradual UVB, and repeated UVC lamp exposure, supplemental to red-blue-white LED chips, also increased phenolics in lettuce cv. Hongyeom; moreover, gene expression of anthocyanin biosynthesis was increased after UV radiation treatment (Lee et al., 2014).

Other secondary metabolite, known to play a major role in plant stress physiology, is ascorbic acid. Bian et al. (2014) noticed that UV light had negative effects on ascorbic acid biosynthesis and accumulation for vegetables cultivated in controlled environments. On the other hand, ascorbate is the main precursor of oxalate in plants. This correlates with reduction of oxalates in silver beets by removing UVB from the light source (Presswood, Hofmann, & Savage, 2012). Stimulating UVA 365 nm LED radiation effect on 3-indolylmethyl glucosinolate accumulation in broccoli (*Brassica oleracea* cv. Italica), compared with violet, blue, green LEDs or fluorescent lamps, was also observed by Rechner, Neugart, Schrener, Wu, and Poehling (2017).

Traditional light sources in controlled environment horticulture (high-pressure sodium, fluorescent, etc.) do not emit much UV light compared with sunlight (Goto et al., 2016). However, the physiological response of green leafy vegetables indicates the potential of UV light application for the improved internal and external quality of the produce. UV effects on secondary metabolites are of the relevant importance for plant immunity and plant/herbivorous interactions, which will be widely reviewed in other chapter.

UVA and UVB Effects on Herbs and Medicinal Plants

In recent years, consumption of fresh herbs or herbal products has increased in daily diets, because of high concentrations of various phytonutrients such as essential oils, phenolic compounds, flavonoids, carotenoids, etc., contributing to the prevention of cardiovascular disease, chronic disease, and some types of cancer. The controlled environment technologies ensure the stable yield and predictable phytochemical quality of herbs and medicinal plants. It also enables the application of specific, controlled environment stresses that can optimize or even enhance the production of pharmacologically active and/or nutritionally important compounds by inducing natural biochemical changes in plants (Dou et al., 2017; Neugart & Schreiner, 2018; Sakalauskaitė et al., 2013, 2012; Zobayed et al., 2005).

The recent development and application of LEDs provides additional methods for targeted regulation of growth and metabolic responses by herbs to optimize productivity and quality under controlled environments (Carvalho et al., 2016; Dou, 2017). In recent years, the effects of red (620–700 nm) and blue (400–490 nm) light and their proportions and combinations with other wavelengths were most researched for the vast of herb species (Amaki et al., 2011; Ceunen, Werbrouck, & Geuns, 2012; Goto et al., 2013; Naznin, Lefsrud, Gravel, & Hao, 2016; Nishimura, Ohyama, Goto, & Inagaki, 2009; Piovene et al., 2015; Sabzalian et al., 2014). However, limited studies have been published regarding the plant response to UV radiation from LEDs. To date, substantial research efforts have covered possible effects of UVB supplementation on herbal plant phytochemicals and biomass production, but experimental studies have so far focused on artificial UVB radiation generated by lamps. It has been shown in several experiments that UVB is positively associated with biomass production in herbal plants (Manaf, Rabie, & Abd El-Aal, 2016; Rai et al., 2011; Sakalauskaitė et al., 2012). However, some studies report decreased biomass with UVB radiation (Nishimura, Ohyama, Goto, Inagaki, & Morota, 2008). UVB light also seems to enhance the accumulation of phytonutrients in herbs. The most widely researched phytonutrients in herbs are essential oils and phenolic compounds (Dou et al., 2017). Light is especially important in regulating the biosynthesis of the volatile compounds that provide the flavor and aroma of leaves, fruits, and flowers. UVB light exposure has been linked to enhanced concentration of essential oils and altered composition of volatile compounds in a range of herb species including sweet flag (*Acorus calamus* L.) (Kumari,

Agrawal, Singh, & Dubey, 2009), Japanese mint (*Mentha arvensis* L. var. Piperascens) (Hikosaka, Ito, & Goto, 2010), lemon balm, sage, lemon catmint (Manukyan, 2013), *Cymbopogon citratus* (Kumari & Agrawal, 2010). UV irradiation for 7 days increased the essential oil content in the upper leaves of Japanese mint plants (*Mentha arvensis* L.) and even UVB or combined UVA and UVB light with white light was more effective than white light alone in increasing L-menthol and limonene concentrations in plants (Hikosaka et al., 2010). Similarly, the content of essential oils in Chinese liquorice (*Glycyrrhiza uralensis*), such as glycyrrhizic acid, liquiritin, liquiritigenin, and isoliquiritigenin in the main root of 3- or 8-month-old plants, was 50%−350% higher under UVB treatment or combined UVA and UVB treatment than the levels in plants irradiated with white light (Sun et al., 2012). Examples of induction of flavor-related compounds by UVB have been reported in basil (*Ocimum basilicum* L.) plants (Behn, Albert, Marx, Noga, & Ulbrich, 2010; Bertoli et al., 2013; Chang, Alderson, & Wright, 2009; Dolzhenko, Bertea, Occhipinti, Bossi, & Maffei, 2010), and these wavebands have been suggested to be required for normal development of oil glands in basil (Ioannidis, Bonner, & Johnson, 2002).

The application of supplemental UVB light could enhance phenolic compounds accumulation in some herb species. The UVB radiation approximately doubled the total phenolic content of rosemary (*Rosmarinus officinalis*) (Luis, Perez, & Gonzalez, 2007), significantly increased in basil (Ghasemzadeh et al., 2016; Mosadegh et al., 2018; Sakalauskaitė et al., 2012, 2013), purple coneflower (*Echinacea purpurea* L.) (Manaf et al., 2016), *Kalanchoe pinnata* (Nascimento et al., 2015), ashwagandha (*Withania somnifera* Dunal) (Takshak & Agrawal, 2014), *Coleus forskohlii* (Takshak & Agrawal, 2015), and others. A further benefit of UVB light exposure is an increase in the concentration of alkaloids in herbs. Takshak and Agrawal (2014) showed that the synthesis of several of the alkaloids in *Withania somnifera* Dunal is simulated by UVB radiation. The synthesis of alkaloids was also induced in *Catharanthus roseus* (Ramani & Chelliah, 2007).

All of these examples show opportunity to control specific processes in herbs with application of relevant UVB light. Such manipulations may be accomplished using narrow-bandwidth light, such as that produced by LED-based light sources. Until nowadays, no experimental studies have been published regarding the implementation of UVB LEDs in greenhouses and other controlled environments for the cultivation of herbal plants. Based on previous research using various types of UVB lamps, targeted low-dosage UVB LED light as a new emerging technology may be used to cultivate herbs enriched with pharmacologically active and/or nutritionally important compounds, also UVB during plant growth leads to more compact and branched plants (Huché-Thélier et al., 2016), which coincide with the consumer preferences for herbs.

In contrast to UVB LEDs, the use of UVA LEDs in horticulture is more advanced (Table 3.1). For herbal plants, the effect of UVA LEDs on basil (*Ocimum basilicum* L.) plant has been the most researched in experimental studies over the last years. Basil is one of the most cultivated commercial herbs, well known as aromatic and medicinal plant used fresh or dried as a drug in traditional medicine and as a favoring agent for foods, confectionary products, and beverages (Sakalauskaite et al., 2012). Basil has antioxidant and antimicrobial activities and ranks among the most important aromatic herbs for its medicinal properties, such as anticancerous, antidiabetic, spasmolytic, carminative, cardioprotective, anthelmintic, and diaphoretic actions (Hussain, Anwar, Sherazi, & Przybylski, 2008; Prakash, Gupta, 2005).

Bantis et al. (2016) reported that even a low dose of supplemental UVA LED light in combination with blue, green, red, far-red light (20% blue light at 400−500 nm + 39% green light at 500−600 nm + 35% red light at 600−700 nm + 5% far-red light at 700−800 nm [red: far red—8.16] + 1% UV light at <400 nm, at $200 \pm 20 \, \mu\text{mol m}^{-2} \text{s}^{-1}$ PPFD and 14 h photoperiod for 28 days) has significant effect on two basil (*Ocimum basilicum* L.) cultivars (Lettuce Leaf, and Red Rubin-mountain Athos hybrid). Basil had greater total biomass, root:shoot ratio was favored, and the highest total phenolic content was obtained. In plant tissues, phenolic compounds are reported to act protectively against UV radiation (Brazaitytė et al., 2015; Iwai, Ohta, Tsuchiya, & Suzuki, 2010; Lee et al., 2014; Li & Kubota, 2009), which might explain the increased total phenolic formation under these light regimes (Bantis et al., 2016). In basil microgreens, the influence of 366, 390, and 402 nm UVA LED wavelengths, supplemental for the basal solid-state lighting system of red, blue, and far-red lights at two UVA irradiation levels (UVA photon flux density of $6.2 \, \mu\text{mol m}^{-2} \text{s}^{-1}$ and of $12.4 \, \mu\text{mol m}^{-2} \text{s}^{-1}$) on the growth and phytochemical contents, was studied (Brazaitytė et al., 2015). The researchers stated that the effect of supplemental UVA irradiation on leaf area and biomass of basil (*Ocimum basilicum* L. cv. Sweet Genovese) depended on the wavelengths and intensity level; the leaf area of basil significantly increased under all supplemental UVA LEDs lighting at the higher intensity level, whereas the most intense growth was under supplemental 402 nm wavelength light and the higher UVA irradiance intensity level. Meanwhile, supplemental UVA LED light at a lower intensity caused a significant or slight increase in fresh weight of basil, also the leaf area of basil significantly increased under supplemental 366 nm wavelength light. Either, the data revealed that almost all supplemental UVA resulted in increased DPPH (2,2-diphenyl-1-picrylhydrazyl) free-radical scavenging activity and total phenolic compounds tend to increase in basil, but supplemental 366 nm at both irradiance level had a harmful effect on the ascorbic acid content (Brazaitytė et al., 2015). The similar effect was observed in another research on basil microgreens and UVA LEDs (Vaštakaitė et al., 2015). Two varieties of basil (*Ocimum basilicum* L.) microgreens purple-leaf cv. Dark Opal and green-leaf cv. Sweet Genovese were grown in greenhouse (14 days, $22/18 \pm 2°C$ day/night temperature, $40 \pm 5\%$ a relative air humidity) during winter season where the main lighting system (HPS lamps and natural daylight) was supplemented with $\sim 13.0 \, \mu\text{mol m}^{-2} \text{s}^{-1}$ flux of UVA 390 nm (total PPFD $\sim 125 \, \mu\text{mol m}^{-2} \text{s}^{-1}$, 16 h photoperiod) for 1 or 7 days before harvest or entire growth period—14 days. The obtained results concluded that the main lighting system supplemented with the UVA LEDs light improves the antioxidant properties in basils and increases total phenol and anthocyanin contents, ascorbic acid concentration, and ABTS (2,2′-azino-bis(3-ethylbenzothiazoline-6-sulfonic acid)) radical scavenging activity, but the effect of

UVA irradiation on phytochemical substances strongly depended on basil variety. Supplemental UVA in the greenhouse was beneficial improving antioxidant properties of green-leaf basils, but this positive impact was not strongly expressed in red-leaf basils (Vaštakaitė et al., 2015).

Park and coauthors (2012) published the first report that demonstrated the positive effect of UVA (380 nm) LED light on the ginsenosides of raw ginseng roots (*Panax ginseng* Meyer). As it is known, ginseng is a perennial herbaceous plant; its roots have been used as herbal medicines for thousands of years. Ginseng extracts contain various compounds, but among them, ginsenosides are considered the most important bioactive ingredients with respect to the pharmacological activities of ginseng. Ginseng roots were exposed to LED light devices with different peak emission wavelengths such as 380, 450, 470, and 660 nm, at 25°C for 7 days in closed compartments. Despite the fact that roots treated with 450 and 470 nm light showed the most pronounced production of ginsenosides (with increases of 64.9% and 74.1%, respectively), the UVA light at 380 nm peaks had also positive effects as compared with the control 4°C-treated roots. The concentrations of protopanaxadiol ginsenosides Rb2 and Rc increased by up to 57% in response to 380 nm (UVA) emission spectrum. In addition, the ratio of protopanaxadiol-type ginsenosides (Rb1, Rb2, Rc, and Rd) to protopanaxatriol-type ginsenosides (Rg1, Rg2, Re, and Rf) in ginseng roots was also changed (was higher than 1.0) by the LED light treatment with 380 nm. The authors concluded that exposure to LED light may allow the production of red ginseng with higher overall levels of pharmacological components and thus a greater commercial value (Park et al., 2012). In another study, Fukuyama, Ohashi-Kaneko, Hirata, Muraoka, and Watanabe (2017) aimed to increase the amount of vinblastine in *Catharanthus roseus* (Apocynaceae) by irradiation of UVA (370 nm) light using an enclosed environmentally controlled room with artificial lighting. *C. roseus* contains more than 130 terpenoid indole alkaloids, of which several, such as vinblastine, vincristine, vinorelbine, and others, are of medicinal interest (Almagro, Fernández-Pérez, & Pedreño, 2015). The regulation system of these alkaloid contents is influenced by environmental conditions, such as light intensity (Fukuyama, Ohashi-Kaneko, & Watanabe, 2015; Liu et al., 2011) and nitrogen content in fertilizer (Gholamhoss, Hemati, Dorodian, & Bashiri-Sadr, 2011; Guo, Chang, Zu, & Tang, 2014). Based on the earlier studies, which indicate that UVA (peak wavelength 370 nm) may induce the increase of vinblastine content (Hirata et al., 1991; 1992; 1993), whereas the monochromatic red light (peak wavelength 660 nm) irradiation increased the growth of *C. roseus* (Fukuyama, Ohashi-Kaneko, Ono, & Watanabe, 2013), to obtain both the high biomass and high vinblastine content in leaves, the plants were grown under red LED light (660 nm) for 28 days and then these plants were irradiated with UVA light (irradiated from UV fluorescent lamps with peak wavelength 370 nm) from 1 week before harvest. At 3 days after treatments, vinblastine content in the leaves increased sharply under UVA supplemented with red light compared with red alone. Also, vinblastine content increased as the UVA intensity was increased from 0 to 10 W m^{-2}. Furthermore, UVA light should be irradiated to the young plants at early stage, because that vinblastine content in the younger leaves was found to be raised as compared with those from the aged leaves. In this experiment, the UV fluorescent lamps with a peak wavelength of 370 nm were used, but this report highlighted that UVA light from LEDs could be applied to cultivate *C. roseus* producing higher content of vinblastine in their leaves using an environmentally controlled plant factory with artificial lighting.

The further study demonstrates that UVA light has a positive effect on polyphenol synthesis, but in this experiment, UVA light was also provided by fluorescent lamps. According to the data of Iwai et al. (2010), artificial illumination with UVA enhanced the content of polyphenols in *Perilla frutescens*, compared with greenhouse-grown plants. The simultaneous artificial light composed of 80% red and 20% blue, followed by irradiation with UVA (80% red laser diode [680 nm] and 20% blue LED [460 nm] for a total PPFD of 360 μmol m^{-2} s^{-1} followed by 2 h of UVA lamp [365 nm] irradiation at 0.13 mW cm^{-2}) for 7 weeks, led to an increase of rosmarinic acid, luteolin glucoside, and caffeic acid in the perilla, compared with greenhouse-grown plants (natural light containing 13.5% UVA and 23.5% blue light, average PPFD was 289 μmol m^{-2} s^{-1}) (Iwai et al., 2010). Biosynthesis of polyphenols is linked to blue light receptors, cryptochromes, and phototropins; therefore, the increase of luteolin-glucoside, caffeic acid, and rosmarinic acid might be closely linked to photoreceptors induction by light. At the same time, UVA has a role in activating the blue light photoreceptors cryptochromes and phototropins (Huché-Thélier et al., 2016) and their involvement in accumulation of polyphenols may be presumed. However, the biosynthesis of polyphenols not only depends on the light spectrum but also strongly depends on plant species. In an experiment involving three common culinary herbs, basil (*Ocimum basilicum* "Genoverser"), arugula (*Eruca sativa*), and bloody dock (*Rumex sanguineus*), grown under supplemental LED light (blue light, maximum peak at 450 nm; blue-violet light, maximum peaks at 420 and 440 nm, which continues to UVA range, i.e., to 380 nm, PAR 300 $\pm$ 10 μmol m^{-2} s^{-1}) in greenhouse, Taulavuori, Pyysalo, Taulavuori, and Julkunen-Tiitto (2018) reported that bloody dock did not respond to supplemental blue or blue-violet light, whereas increased phytochemicals in both, basil and arugula, with couple of species-specific responses were found. Blue and blue-violet light increased production of phenolic acids in basil, whereas arugula increased synthesis of flavonoid compounds. In addition, the phenolic acids in basil were independent of quality of blue light (blue vs. blue-violet); on the contrary, blue-violet light that contained also small proportion of UVA was even more efficient in stimulation of flavonoid synthesis in arugula, indicating the role of flavonoids in protection against UVA. Flavonoids also are major blue light—absorbing pigments in a range of 400—430 nm (Gitelson, Chivkunova, Zhigalova, & Solovchenko, 2017), close to UVA spectrum (i.e., 315—400 nm).

Modifications of the spectral environment by LED supplemental lighting on the background of natural daylight can produce relevant effects on postcultivation performance of plants, which is sensitive to chilling injuries (Jensen, Clausen, & Kjaer, 2018). In basils grown under greenhouse conditions using four different supplemental LED light treatments (80%Red/20%Blue, 80%Red/20%Blue + UVA, 40%

Red/60%Blue, and 80%Red/20%Green), the addition of UVA to red and blue light showed significant effect on the chilling tolerance and shelf life performance. The tolerance toward chilling was highest for the 80%Red/20%Green treatment, the other ones were in the order 80%Red/20%Blue + UVA > 80%Red/20% Blue > 40%Red/60%Blue; this trend remained throughout the shelf life simulation.

Several studies have revealed that UVC light can also have remarkable and promoting effects on the accumulation of phytochemicals and antioxidants activity in various plants (Alothman, Bhat, & Karim, 2009; Dogu-Baykut, Gurbuz Gunes, & Decker, 2014; Liu, Cai, Lu, Han, & Ying, 2012; Moon, Mistry, Kim, & Panduran-gan, 2017; Perkins-Veazie, Collins, & Howard, 2008). The recent study (Moon et al., 2017) revealed that UVC radiation (UVC lamp) enhanced the accumulation of bioactive compounds such as total flavonoids, total phenolic, and terpenoid indole alkaloids and also induced the antioxidant capacity in cambium meristematic cells from *Catharanthus roseus* (L.) G. Don. The authors concluded that the UVC-treated *C. roseus* cambium meristematic cells are a potential candidate to be used as natural antioxidant and anticancer agent, and treatment of UVC can be used as an efficient method to enhance bioactive compounds. The effect of nondeleterious doses of UVC light has been exploited successfully to control postharvest diseases, thus extending shelf life of horticulture crops (Urban et al., 2018). Therefore, future studies should focus on the effect of UVC radiation from LEDs to stimulate the production of health-promoting phytochemicals, to extent shelf life of horticulture crops and to stimulate mechanisms of adaptation to biotic and abiotic stresses. Much less is known about UVC light even though it seems that carefully selected doses would have a lot of potential for triggering useful effects in plants in less time than UVB light (Urban, Charles, de Miranda, & Aarrouf, 2016).

Given the novelty of legal commercial cannabis production, more and more professional medicinal cannabis (*Cannabis sativa* L.) producers are moving to CEA, with possibility to control temperature, humidity, light, CO_2 concentration, etc. Medicinal cannabis producer seeks continuous and uniform yield, but production of specific cannabinoid compound or ratio between the different cannabinoids, especially in floral buds, is of primary interest. Light, its quality, intensity, and photoperiod play a great role in a successful growth protocol of medicinal cannabis. Already the very early studies demonstrated that in different light environments, it is possible to manipulate the cannabinoid content of *Cannabis sativa* L. (Mahlberg & Hemphill, 1983). However, over the past years, there are relatively few scientific studies on cannabis production data; also it is likely that producers have not yet determined the optimal light for their specific cultivars and production methods or commercial cannabis producers typically guard their production strategies (Hawley, Graham, Stasiak, & Dixon, 2018). With reference to recent research, the medicinal cannabis cultivation can strongly benefit from LED lighting with respect to growth, development, and biochemical profile (Bantis et al., 2018). A recently published paper (Magagnini, Grassi, & Kotiranta, 2018) suggests that an optimized light spectrum improves the value and quality of cannabis and LED technology has significant differences in growth and cannabinoid profile compared with the traditional HPS light source. Cannabis plants under HPS (300–400 nm—1%, 400–500 nm—8%, 500–600 nm—68%, 600–700 nm—21%, 700–800 nm—3%, 400–700 PAR—96%; R:FR [650–670 nm/720–740 nm]—2.80, B:G [420–490 nm/500–570 nm]—0.29, B:R—0.10) treatment were taller and had more flower dry weight than those under LED lamp (Valoya, Finland) treatments: AP673L (300–400 nm—0%, 400–500 nm—14%, 500–600 nm—20%, 600–700 nm—59%, 700–800 nm—7%, 400–700 PAR—93%; R:FR (650–670 nm/720–740 nm)—6.07, B:G (420–490 nm/500–570 nm)—1.76, B:R—0.26), and NS1 (300–400 nm—2%, 400–500 nm—24%, 500–600 nm—37%, 600–700 nm—33%, 700–800 nm—4%, 400–700 PAR—94%; R:FR (650–670 nm/720–740 nm)—10.05, B:G (420–490 nm/500–570 nm)—0.74, B:R—0.80). However, HPS resulted in a significant decline of Δ-9tetrahydrocannabinol (THC) concentration in flowers compared to both LED treatments. In general, the authors of this research stated that the highest cannabigerol (CBG) and THC concentrations were measured in the treatment, which had the highest portion of blue and UVA wavelengths in the spectrum compared with the other treatments. The complex mechanisms mediated by the UVA and blue wavelengths may act synergistically to induce CBG accumulation in cannabis flowers, whereas CBG is the precursor of other cannabinoids (Magagnini et al., 2018). Recently published paper from Canada also concluded that cannabis plants grown under supplemental subcanopy lighting (SCL) can increase bud yield and modify cannabinoid and terpene profiles (Hawley et al., 2018). LED red/blue (Red-Blue) and red-green-blue (RGB) supplemental SCL treatments significantly increased dry bud yield of cannabis more than the control (plants were grown with no supplemental SCL) treatment. The research results suggest that Red-Blue SCL yielded a consistently more stable metabolome profile between the upper and lower canopy than RGB or control treatment with the greatest impact on upregulating metabolites. Overall, RGB SCL had the greatest impact on modifying terpene content, and Red-Blue produced a more homogenous bud cannabinoid and terpene profile throughout the canopy (Hawley et al., 2018). Study regarding treatment of red, green, or blue LEDs emitting light on cannabis sprouts leads to conclusion that source of light (sunlight or LEDs irradiation) and color of LEDs influences the growth, phytochemical compounds content, antioxidant activity, and protein concentration in cannabis sprouts (Livadariu, Raiciu, Maximilian, & Căpitanu, 2018). The illumination with blue light LEDs induced the superior rate and fresh weight of cannabis sprouts, whereas the lowest value was for sunlight treatment. Also the content of polyphenols, flavonoids, and protein was significantly higher by treatment with blue light LEDs, whereas green light LEDs improved the antioxidant capacity. Another study in relation to cannabis growth and development, conducted in a controlled indoor growing facility, indicates that plants grown under red (630–680 nm) and blue (420–490 nm) LED light spectrum was lower and had a smaller leaf area compared with a white light source (Lalge, Cerny, Trojan, & Vyhnanek, 2017). The authors concluded that white light emitting a full spectrum of light had more significant effect on cannabis plant growth and development than red blue light.

The Effects of UVA and UVB Irradiation on Fruiting Greenhouse Crops

In greenhouses and controlled environment agricultural systems, it is possible to produce vegetables whole year despite the cultivation season (Bian et al., 2014; Mariz-Ponte et al., 2018). However, even using artificial lighting, vegetables produced off-season often have diminished organoleptic properties and chemical composition, compared with those produced in open fields (Mariz-Ponte et al., 2018; Muñoz et al., 2007). Plants, grown in controlled environment horticultural systems (both greenhouses and plant factories), are not exposed to natural doses of UV radiation, which is known to have significant impact on fruit production, sensorial properties, and phytochemical contents (Carvalho et al., 2016; Mariz-Ponte et al., 2018). Therefore, UV LED lighting strategies, optimized for fruiting vegetables (tomatoes, cucumbers, sweet peppers, eggplants, etc.), could be beneficial both for plant growth and fruit quality (Table 3.1).

UV light, similar to blue light, was shown to affect positively tomato seedling growth. Khoshimkhujaev, Kwon, Park, Choi, and Lee (2014) explored that 376 nm UVA LED irradiation, supplemental for red 658 nm LEDs, had beneficial effect on the growth and development of tomato (*Lycopersicon esculentum* L. cv. Superdoterang) seedlings. Similar to blue light effect, tomato seedlings, illuminated with UVA, became more compact, the growth of plant organs was balanced, the leaf area was increased, and the total plant fresh and dry weights were also enhanced (Khoshimkhujaev et al., 2014). Similar effects were obtained by other authors. In the study of Brazaitytė et al. (2010), UVA 380 nm LEDs, supplemental for the main red, blue, and far-red LED lighting, resulted in enhanced growth of tomato (*Lycopersicon esculentum* L. Raissa F1) seedlings: it increased fresh and dry plant weight, contents of chlorophylls in the leaves, hypocotyl diameter, leaf area. When seedlings were transplanted into the greenhouse with no artificial lighting, the "aftereffect" of UVA LED light on the contents of the photosynthetic pigments was observed for about 2 weeks, but had no significant effect on early tomato yield (Brazaitytė, Kasiulevičiūtė-Bonakerė et al., 2009; Brazaitytė, Kazėnas et al., 2009). Yang et al. (2018) compared white, red, red/blue, blue, and purple ($\sim$400 nm) LED lighting effects on tomato cv. Liao Yuan Duo Li seedlings. Purple light, similar to blue light, resulted in more compact tomato growth; lower contents of photosynthetic pigments in leaves reduced photosynthetic efficiency but increased antioxidant contents in leaves, which meets the symptoms of photoprotection (Yang et al., 2018). Kim, Park, Park, Lee, and Oh (2014) investigated effects of various wavelengths of LED light and concluded that the increase in the level of phenolics and antioxidant capacity in tomato cv. Cuty leaves reflects the environmental resistance and, possibly, the further transplanting performance (Kim et al., 2014).

Mariz-Ponte et al. (2018) explored that both UVA and UVB lights, supplemental for white fluorescent light, were efficient in the flowering/fruiting process synchronization in tomato cv. Micro Tom with minimal impacts on the vegetative part.

Numerous studies also confirm UV light effects on phytochemical contents in tomato fruits. UV stimulates accumulation of lycopene, as well as other carotenes and xanthophylls (Lazzeri et al., 2012; Llorente et al., 2015); the levels of caffeic acid, ferulic acid, and p-coumaric acid are higher in plants exposed to higher UVB doses compared with plants that are not; however, phenolic acids are not affected by UVB (Neugart & Schreiner, 2018).

Similar trends of UV light on plant performance, morphology, and phenolic compound production were obtained with pepper plants. Rodríguez-Calzada et al. (2019) investigating UVB effects on chili pepper (*Capsicum annuum*, cv. Coronel) concluded that UVB induced a reduction in stem length, stem dry weight, and number of floral primordia and increased epidermal flavonoids (specifically chlorogenic acid and apigenin 8-C-hexoside) levels in leaves. León-Chan et al. (2017) in bell pepper cv. Cannon, exposed to UVB, detected degradation of chlorophyll in leaves and the higher accumulation of carotenoids, chlorogenic acid, and the flavonoid apigenin-7-O-glucoside. However, the study of Bagdonavičienė et al. (2015) explored that low flux of UVA 380 nm LED light, supplemental for red, blue, and far-red LEDs, had no beneficial effect on sweet pepper (*Capsicum annum* L. cv. Reda transplants).

Cucumber seedlings were found to be more sensitive to UV light, compared with tomato. Experiments in fields using UV blocking filters (Krizek, Mirecki, & Britz, 1997) showed that UVB and UVA light reduced biomass of leaves, stems, and roots and diminished stem height and leaf area; however, UVB light had more pronounced effect compared with UVA. Peng, Qiang, Yunyun, Hongjun, and Weijie (2017) investigated the impact of UVB light, provided by fluorescent tubes, supplemental for high-pressure sodium lamp lighting in greenhouse on cucumber (*Cucumis sativus* L. cv. Chinese long 9930) growth. The results showed that UVB radiation effectively inhibited the elongation of cucumber and decreased soluble protein content in leaves. 3.33 μmol m^{-2} s^{-1} UVB promoted stem diameter growth, soluble sugar content, total ascorbic acid, and the superoxide dismutase, peroxidase, and catalase activities in cucumber leaves. However, this UVB dose had no influence on the net photosynthetic rate of cucumber leaves; therefore, it is beneficial for cucumber cultivation in controlled environment systems (Peng et al., 2017). Brazaitytė, Kasiulevičiūtė-Bonakerė et al. (2009) and Brazaitytė, Kazėnas et al. (2009) investigated the impact of UVA LEDs (380 nm), supplemental for red, blue, and far-red LED light, on cucumber hybrid Mandy F1 seedling growth and yield after transplanting. The results explored that supplemental UVA LED light decreased growth and development of cucumber seedlings. After transplanting in nonilluminated greenhouse, the inhibiting impact of UVA light remained visible for 2 weeks and delayed the flowering and first harvest but had no significant effect on cucumber yield (Brazaitytė, Kasiulevičiūtė-Bonakerė et al., 2009; Brazaitytė, Kazėnas et al., 2009).

In case of cucumber, UVA or UVB light had no significantly positive impact. However, UV deficiency in lighting spectrum might result in serious plant damages, if seedlings, cultivated under UV-deficient spectrum in controlled environment agricultural system, would be directly transplanted to natural illumination environment. Therefore, the period of adaptation, protecting seedlings from direct solar light exposure, is necessary.

Summary

UVB and UVA lights were shown to impact growth, photosynthesis, secondary plant metabolites, and the plant-insect interaction in important agricultural and horticultural crops (Davis & Burns, 2016), such as lettuce, herbs and medicinal plants, tomatoes, and other fruiting greenhouse crops. An increase in understanding of basic plant physiology, UV response, and tolerance mechanisms can be gained from studies reviewed. The emerging possibilities to apply specific wavelengths of UV LEDs for plant lighting, seeking for anticipated effects on growth, morphology, and phytochemical quality, motivate for further research and applications. Properly balanced UV LED parameters (wavelength, intensity, timing, and duration of exposure), consistent with other environmental factors, might create plant species-specific eustress conditions and therefore be useful in agroindustry (Hideg et al., 2013; Mariz-Ponte et al., 2018). One of the main recent challenges for the increased use of UV LEDs in horticulture includes the need to develop biological and technical outlines for efficient and profitable production outcomes.

Acknowledgments

This research was funded by a grant (No. S-MIP-17-23) from the Research Council of Lithuania.

References

Agati, G., Biricolti, S., Guidi, L., Ferrini, F., Fini, A., & Tattini, M. (2011). The biosynthesis of flavonoids is enhanced similarly by UV radiation and root zone salinity in *L. vulgare* leaves. *Journal of Plant Physiology, 168*, 204—212.

Agati, G., Brunetti, C., Diferdinando, M., Ferrini, F., Pollastri, S., & Attini, M. (2013). Functional roles of flavonoids in photoprotection: New evidence, lessons from the past. *Plant Physiology and Biochemistry, 72*, 35—45.

Almagro, L., Fernández-Pérez, F., & Pedreño, M. A. (2015). Indole alkaloids from *Catharanthus roseus*: Bioproduction and their effect on human health. *Molecules, 20*, 2973—3000.

Alothman, M., Bhat, R., & Karim, A. A. (2009). UV radiation-induced changes of antioxidant capacity of fresh-cut tropical fruits. *Innovative Food Science & Emerging Technologies, 10*, 512—516.

Amaki, W., Yamazaki, N., Ichimura, M., & Watanabe, H. (2011). Effects of light quality on the growth and essential oil content in sweet basil. *Acta Horticulturae, 907*, 91—94.

Bagdonavičienė, A., Brazaitytė, A., Jankauskienė, J., Viršilė, A., Sirtautas, A., Samuolienė, G., et al. (2015). The effect of light-emitting diodes illumination spectra on growth of sweet pepper (*Capsicum annum* L.) transplants. In *Proceedings of 25th congress of the nordic association of agricultural scientists "nordic view to sustainable rural development", Riga, Latvia, 54*.

Ballare, C. L., Caldwell, M. M., Flint, S. D., Robinson, S. A., & Bornman, J. F. (2011). Effects of solar ultraviolet radiation on terrestrial ecosystems. Patterns, mechanisms, and interactions with climate change. *Photochemical and Photobiological Sciences, 10*, 226–241.

Bantis, F., Ouzounis, T., & Radoglou, K. (2016). Artificial LED lighting enhances growth characteristics and total phenolic content of Ocimum basilicum, but variably affects transplant success. *Scientia Horticulturae, 198*, 277–283.

Bantis, F., Smirnakou, S., Ouzounis, T., Athanasios Koukounaras, A., Ntagkas, N., & Radoglou, K. (2018). Current status and recent achievements in the field of horticulture with the use of light-emitting diodes (LEDs). *Scientia Horticulturae, 235*, 437–451.

Baroniya, S. S., Kataria, S., Pandey, G. P., & Guruprasad, K. N. (2014). Growth, photosynthesis and nitrogen metabolism in soybean varieties after exclusion of the UV-B and UV-A/B components of solar radiation. *Crop J, 2*, 388–397.

Bassman, J. H. (2004). Ecosystem consequences of enhanced solar ultraviolet radiation: Secondary plant metabolites as mediators of multiple trophic interactions in terrestrial plant communities. *Photochemistry and Photobiology, 79*(5), 382–398.

Behn, H., Albert, A., Marx, F., Noga, G., & Ulbrich, A. (2010). Ultraviolet-B and photosynthetically active radiation interactively affect yield and pattern of monoterpenes in leaves of peppermint (*Mentha × piperita* L.). *Journal of Agricultural and Food Chemistry, 58*, 7361–7367.

Bertoli, A., Lucchesini, M., Menuali-Sodi, A., Leonardi, M., Doveri, S., Magnabosco, A., et al. (2013). Aroma characterization and UV elicitation of purple basil from different plant tissue cultures. *Food Chemistry, 141*, 776–787.

Bian, Z. H., Yang, Q. C., & Liu, W. K. (2014). Effects of light quality on the accumulation of phytochemicals in vegetables produced in controlled environments: A review. *Journal of the Science of Food and Agriculture, 95*(5), 869–877.

Bilger, W., Rolland, M., & Nybakken, L. (2007). UV screening in higher plants induced by low temperature in the absence of UV-B radiation. *Photochemical and Photobiological Sciences, 6*(2), 190–195.

Brazaitytė, A., Duchovskis, P., Urbonavičiūtė, A., Samuolienė, G., Jankauskienė, J., Kasiulevičiūtė-Bonakerė, A., et al. (2009). The effect of light-emitting diodes lighting on cucumber transplants and after-effect on yield. *Zemdirbyste, 96*(3), 102–118.

Brazaitytė, A., Duchovskis, P., Urbonavičiūtė, A., Samuolienė, G., Jankauskienė, J., Kazėnas, V., et al. (2009). After-effect of light-emitting diodes lighting on tomato growth and yield in greenhouse. *Sodininkystė ir Daržininkystė, 28*(1), 115–126.

Brazaitytė, A., Duchovskis, P., Urbonavičiūtė, A., Samuolienė, G., Jankauskienė, J., Sakalauskaitė, J., et al. (2010). The effect of light-emitting diodes lighting on the growth of tomato transplants. *Zemdirbyste, 97*(2), 89–98.

Brazaitytė, A., Viršilė, A., Jankauskienė, J., Sakalauskienė, S., Samuolienė, G., Sirtautas, R., et al. (2015). Effect of supplemental UV-A irradiation in solid-state lighting on the growth and phytochemical content of microgreens. *International Agrophysics, 29*, 13–22.

Brazaitytė, A., Viršilė, A., Samuolienė, G., Jankauskienė, J., Sakalauskienė, S., Sirtautas, R., et al. (2016). Light quality: Growth and nutritional value of microgreens under indoor and greenhouse conditions. *Acta Horticulturae, 1134*, 277–284.

Britz, S. J., Caldwell, C. R., Mirecki, R., Slusser, J., & Gao, W. (2005). Effect of supplemental ultraviolet radiation on the concentration of phytonutrients in green and red leaf lettuce (*Lactuca sativa*) cultivars. In *Proceedings of SPIE* (Vol. 5886, pp. 1−8).

Bugbee, B. (2016). Toward an optimal spectral quality for plant growth and development: The importance of radiation capture. *Acta Horticulturae, 1134*, 1−12.

Bugbee, B. (2017). Economics of LED lighting. In S. D. Gupta (Ed.), *Light emitting diodes for agriculture: Smart lighting* (p. 334). Springer Nature Singapore Ltd. https://doi.org/10.1007/978-981-10-5807-3.

Caldwell, C. R., & Britz, S. J. (2006). Effect of supplemental ultraviolet radiation on the carotenoid and chlorophyll composition of greenhouse-grown leaf lettuce (*Lactuca sativa* L.) cultivars. *Journal of Food Composition and Analysis, 19*, 637−644.

de Carbonnel, M., Davis, P. A., Rob, M., Roelfsema, G., Inoue, S., Schepens, I., et al. (2010). The *Arabidopsis* phytochrome kinase substrate 2 protein is a phototropin signalling element that regulates leaf flattening and leaf positioning. *Plant Physiology, 152*, 1391−1405.

Carvalho, S. D., Schwieterman, M. L., Abrahan, C. E., Colquhoun, T. A., & Folta, K. M. (2016). Light quality dependent changes in morphology, antioxidant capacity, and volatile production in sweet basil (*Ocimum basilicum*). *Frontiers of Plant Science, 7*, 1328.

Castilla, N., & Hernandez, J. (2006). Greenhouse technological packages for high-quality crop production. *Acta Horticulturae, 761*, 285−297.

Ceunen, S., Werbrouck, S., & Geuns, J. M. C. (2012). Stimulation of steviol glycoside accumulation in *Stevia rebaudiana* by red LED light. *Journal of Plant Physiology, 169*, 749−752.

Chang, X., Alderson, P. G., & Wright, C. J. (2009). Enhanced UV-B radiation alters basil (*Ocimum basilicum* L.) growth and stimulates the synthesis of volatile oils. *Journal of Horticulture and Forestry, 1*, 27−31.

Christie, J. M., Arvai, A. S., Baxter, K. J., Heilmann, M., Pratt, A. J., O'Hara, A., et al. (2012). Plant UVR8 photoreceptor senses UV-B by tryptophan-mediated disruption of cross-dimer salt bridges. *Science, 335*(6075), 1492−1496.

Cockell, M. M., & Knowland, J. (1999). Ultraviolet radiation screening compounds. *Biological Reviews, 74*, 311−345.

Czégény, G., Mátai, A., & Hideg, E. (2016). UV-B effects on leaves − oxidative stress and acclimation in controlled environments. *Plant Science, 248*, 57−63.

Davis, P. A., & Burns, C. (2016). Photobiology in protected horticulture. *Food and Energy Security, 5*(4), 223−238.

Despommier, D. (2010). *The vertical farm: Feeding the world in the 21st century*. New York, NY, USA: Macmillan, 2010.

Dogu-Baykut, E., Gurbuz Gunes, G., & Decker, E. A. (2014). Impact of shortwave ultraviolet (UV-C) radiation on the antioxidant activity of thyme (*Thymus vulgaris* L.). *Food Chemistry, 157*, 167−173.

Dolzhenko, Y., Bertea, C., Occhipinti, A., Bossi, S., & Maffei, M. (2010). UV-B modulates the interplay between terpenoids and flavonoids in peppermint (*Mentha* × *piperita* L.). *Journal of Photochemistry and Photobiology B, 100*, 67−75.

Dou, H., Niu, G., Gu, M., & Masabni, J. G. (2017). Effects of light quality on growth and phytonutrient accumulation of herbs under controlled environments. *Horticulturae, 3*(2), 36.

D'Souza, C., Yuk, H. G., Khoo, G. H., & Zhou, W. (2015). Application of light-emitting diodes in food production, postharvest preservation, and microbiological food safety. *Comprehensive Reviews in Food Science and Food Safety, 14*, 719−740.

Ebisawa, M., Shoji, K., Kato, M., Shimomura, K., Goto, F., & Yoshihara, T. (2008). Supplementary ultraviolet radiation B together with blue light at night increased quercetin content and flavonol synthase gene expression in leaf lettuce (*Lactuca sativa* L.). *Environment Control in Biology, 46*, 1−11.

Escobar-Bravo, R., Klinkhamer, P. G. L., & Leiss, K. A. (2017). Interactive effects of UV-B light with abiotic factors on plant growth and chemistry, and their consequences for defense against arthropod herbivores. *Frontiers of Plant Science, 8*, 278.

Fina, J., Casadevall, R., AbdElgawad, H., Prinsen, E., Markakis, M. N., Beemster, G. T. S., et al. (2017). UV-B inhibits leaf growth through changes in growth regulating factors and gibberellin levels. *Plant Physiology, 174*, 1110−1126.

Fukuyama, T., Ohashi-Kaneko, K., Hirata, K., Muraoka, M., & Watanabe, H. (2017). Effects of ultraviolet A supplemented with red light irradiation on vinblastine production in *Catharanthus roseus*. *Environment Control in Biology, 55*, 65−69.

Fukuyama, T., Ohashi-Kaneko, K., Ono, E., & Watanabe, H. (2013). Growth and alkaloid yields of *Catharanthus roseus* (L.) G. Don cultured under red and blue LEDs. *Journal of Science and Technology of Agriculture, 25*, 175−182.

Fukuyama, T., Ohashi-Kaneko, K., & Watanabe, H. (2015). Estimation of optimal red light intensity for production of the pharmaceutical drug components, vindoline and catharanthine, contained in *Catharanthus roseus* (L.) G. Don. *Environment Control in Biology, 53*, 217−220.

Gardner, G., Lin, C., Tobin, E. M., Loehrer, H., & Brinkman, D. (2009). Photobiological properties of the inhibition of etiolated *Arabidopsis* seedling growth by ultraviolet-B irradiation. *Plant, Cell and Environment, 32*, 1573−1583.

Ghasemzadeh, A., Ashkani, S., Baghdadi, A., Pazoki, A., Jaafar, H. Z., & Rahmat, A. (2016). Improvement in flavonoids and phenolic acids production and pharmaceutical quality of sweet basil (*Ocimum basilicum* L.) by ultraviolet-B irradiation. *Molecules, 21*(9), 1203−1218.

Gholamhoss, Z., Hemati, K., Dorodian, H., & Bashiri-Sadr, Z. (2011). Effect of nitrogen fertilizer on yield and amount of alkaloids in periwinkle and determination of vinblastine and vincristine by HPLC and TLC. *Plant Sciences Research, 3*, 4−9.

Gitelson, A., Chivkunova, O., Zhigalova, T., & Solovchenko, A. (2017). In situ optical properties of foliar flavonoids: Implication for non-destructive estimation of flavonoid content. *Journal of Plant Physiology, 218*, 257−264.

Goto, E., Hayashi, K., Furuyama, S., Hikosaka, S., & Ishigami, Y. (2016). Effect of UV light on phytochemical accumulation and expression of anthocyanin biosynthesis genes in red leaf lettuce. *Acta Horticulturae, 1134*, 179−185.

Goto, E., Matsumoto, H., Ishigami, Y., Hikosaka, S., Fujiwara, K., & Yano, A. (2013). Measurements of the photosynthetic rates in vegetables under various qualities of light from light-emitting diodes. *Acta Horticulturae, 1037*, 261−268.

Guo, X. R., Chang, B. W., Zu, Y. G., & Tang, Z. H. (2014). The impacts of increased nitrate supply on *Catharanthus roseus* growth and alkaloid accumulations under ultraviolet-B stress. *Journal of Plant Interactions, 9*, 640−646.

Hawley, D., Graham, T., Stasiak, M., & Dixon, M. (2018). Improving cannabis bud quality and yield with subcanopy lighting. *HortScience, 53*(11), 1593−1599.

Hayes, S., Velanis, C. N., Jenkins, G. I., & Franklin, K. A. (2014). UV-B detected by the UVR8 photoreceptor antagonizes auxin signaling and plant shade avoidance. *Proceedings of the National Academy of Sciences of the United States of America, 111*, 11894−11899.

Hectors, K., van Oevelen, S., Guisez, Y., Prinsen, E., & Jansen, M. A. K. (2012). The phytohormone auxin is a component of the regulatory system that controls UV mediated accumulation of flavonoids and UV-induced morphogenesis. *Physiologia Plantarum, 145,* 594—603.

Hideg, E., Jansen, M. A. K., & Strid, A. (2013). UV-B exposure, ROS, and stress: Inseparable companions or loosely linked associates? *Trends in Plant Science, 18,* 107—115.

Hikosaka, S., Ito, K., & Goto, E. (2010). Effects of ultraviolet light on growth, essential oil concentration, and total antioxidant capacity of Japanese mint. *Environment Control in Biology, 48,* 185—190.

Hirata, K., Asada, M., Yatani, E., Miyamoto, K., & Miura, Y. (1993). Effects of nearultraviolet light on alkaloid production in *Catharanthus roseus* plants. *Planta Medica, 59,* 46—50.

Hirata, K., Horiuchi, M., Ando, T., Asada, M., Miyamoto, K., & Miura, Y. (1991). Effect of near-ultraviolet light on alkaloid production in multiple shoot cultures of *Catharanthus roseus. Planta Medica, 57,* 499—500.

Hirata, K., Horiuchi, M., Asada, M., Ando, T., Miyamoto, K., & Miura, Y. (1992). Stimulation of dimeric alkaloid production by near-ultraviolet light in multiple shoot cultures of *Catharanthus roseus. Journal of Fermentation and Bioengineering, 74,* 222—225.

Hoffmann, A. M., Noga, G., & Hunsche, M. (2015). High blue light improves acclimation and photosynthetic recovery of pepper plants exposed to UV stress. *Environmental and Experimental Botany, 109,* 254—263.

Hollósy, F. (2002). Effects of ultraviolet radiation on plant cells. *Micron, 33,* 179—197.

Huché-Thélier, L., Crespel, L., Le Gourrierec, J., Morel, P., Sakr, S., & Leduc, N. (2016). Light signaling and plant responses to blue and UV radiations—perspectives for applications in horticulture. *Environmental and Experimental Botany, 121,* 22—38.

Hussain, A. I., Anwar, F., Sherazi, S. T. H., & Przybylski, R. (2008). Chemical composition, antioxidant and antimicrobial activities of basil (*Ocimum basilicum*) essential oils depend on seasonal variations. *Food Chemistry, 108,* 986—995.

Ioannidis, D., Bonner, L., & Johnson, C. B. (2002). UV-B is required for normal development of oil glands in *Ocimum basilicum* L. (sweet basil). *Annals of Botany, 90,* 453—460.

Iwai, M., Ohta, M., Tsuchiya, H., & Suzuki, T. (2010). Enhanced accumulation of caffeic acid, rosmarinic acid and luteolin-glucoside in red perilla cultivated under red diode laser and blue LED illumination followed by UV-A irradiation. *Journal of Functional Foods, 2*(1), 66—70.

Jansen, M. A. K., & Bornman, J. F. (2012). UV-B radiation: From generic stressor to specific regulator. *Physiologia Plantarum, 145,* 501—504.

Jansen, M. A., & Van Den Noort, R. E. (2000). Ultraviolet-B radiation induces complex alterations in stomatal behaviour. *Physiologia Plantarum, 110,* 189—194.

Jenkins, G. I. (2009). Signal transduction in responses to UV-B radiation. *Annual Review of Plant Biology, 60,* 407—431.

Jenkins, G. (2017). Photomorphogenic responses to ultraviolet-B light. *Plant, Cell and Environment, 40*(11), 2544—2557.

Jensen, N. B., Clausen, M. R., & Kjaer, K. H. (2018). Spectral quality of supplemental LED grow light permanently alters stomatal functioning and chilling tolerance in basil (*Ocimum basilicum* L.). *Scientia Horticulturae, 227,* 38—47.

Jordan, B. R. (2002). Molecular response of plant cells to UV-B stress. *Functional Plant Biology, 29,* 909—916.

Kakani, V. G., Reddy, K. R., Zhao, D., & Sailaja, K. (2003). Field crop responses to ultraviolet-B radiation: A review. *Agricultural and Forest Meteorology, 120*, 191–218.

Kataria, S., Jajoo, A., & Guruprasad, K. N. (2014). Impact of increasing Ultraviolet-B (UV-B) radiation on photosynthetic processes. *Journal of Photochemistry and Photobiology B, 137*, 55–66.

Khoshimkhujaev, B., Kwon, J. K., Park, K. S., Choi, H. G., & Lee, S. Y. (2014). Effect of monochromatic UV-A LED irradiation on the growth of tomato seedlings. *Horticulture, Environment, and Biotechnology, 55*(4), 287–292.

Kim, E. Y., Park, S. A., Park, B. J., Lee, Y., & Oh, M. M. (2014). Growth and antioxidant phenolic compounds in cherry tomato seedlings grown under monochromatic light-emitting diodes. *Horticulture, Environment, and Biotechnology, 55*(6), 506–513.

Kneissl, M., & Rass, J. (2016). III-Nitride ultraviolet emitters. In *Springer series in materials science*. Cham: Springer International Publishing. https://doi.org/10.1007/978-3-319-24100-5.

Kozai, T. (2013). Resource use efficiency of closed plant production system with artificial light: Concept, estimation and application to plant factory. *Proceedings of the Japan Academy Series B Physical and Biological Sciences, 89*, 447–461.

Kozai, T., Niu, G., & Takagaki, M. (2015). *Plant factory: An indoor vertical farming system for efficient quality food production*. San Diego, CA, USA: Academic Press.

Krizek, D. T., Britz, S. J., & Mirecki, R. M. (1998). Inhibitory effects of ambient levels of solar UV-A and UV-B radiation on growth of cv. New Red Fire lettuce. *Physiologia Plantarum, 103*, 1–7.

Krizek, D. T., Mirecki, R. M., & Britz, S. J. (1997). Inhibitory effects of ambient levels of solar UV-A and UV-B radiation on growth of cucumber. *Physiologia Plantarum, 100*(4), 886–893.

Kumari, R., & Agrawal, S. B. (2010). Supplemental UV-B induced changes in leaf morphology, physiology and secondary metabolites of an Indian aromatic plant *Cymbopogon citratus* (D.C.) Staph under natural field conditions. *The International Journal of Environmental Studies, 67*, 655–675.

Kumari, R., Agrawal, S. B., Singh, S., & Dubey, N. K. (2009). Supplemental ultraviolet-B induced changes in essential oil composition and total phenolics of *Acorus calamus* L. (sweet flag). *Ecotoxicology and Environmental Safety, 72*, 2013–2019.

Lalge, A., Cerny, P., Trojan, V., & Vyhnanek, T. (2017). The effect of red, blue and white light on the growth and development of *Cannabis sativa* L. MendelNet 2017. In *Proceedings of 24th international PhD students conference* (Vol. 24, pp. 646–651). Czech Republic: Mendel University in Brno, ISBN 978-80-7509-529-9.

Lazzeri, V., Calvenzani, V., Petroni, K., Tonelli, C., Castagna, A., & Ranieri, A. (2012). Carotenoid profiling and biosynthetic gene expression in flesh and peel of wild-type and hp-1 tomato fruit under UV-B depletion. *Journal of Agricultural and Food Chemistry, 60*, 4960–4969.

Lee, M. J., Son, J. E., & Oh, M. M. (2014). Growth and phenolic compounds of *Lactuca sativa* L. grown in a closed-type plant production system with UV-A, -B, or -C lamp. *Journal of the Science of Food and Agriculture, 94*(2), 197–204.

Lefsrud, M. G., Kopsell, D. A., & Sams, C. E. (2008). Irradiance from distinct wavelength light-emitting diodes affect secondary metabolites in kale. *HortScience, 43*(7), 2243–2244.

León-Chan, R. G., López-Meyer, M., Osuna-Enciso, T., Sañudo-Barajas, J. A., Heredia, J. B., & León-Félix, J. (2017). Low temperature and ultraviolet-B radiation affect chlorophyll content and induce the accumulation of UV-B-absorbing and antioxidant compounds in bell pepper (*Capsicum annuum*) plants. *Environmental and Experimental Botany, 139*, 143—151.

Li, Q., & Kubota, C. (2009). Effects of supplemental light quality on growth and phytochemicals of baby leaf lettuce. *Environmental and Experimental Botany, 67*, 59—64.

Liu, C. H., Cai, L. Y., Lu, X. Y., Han, X. X., & Ying, T. J. (2012). Effect of postharvest UV-C irradiation on phenolic compound content and antioxidant activity of tomato fruit during storage. *Journal of Integrative Agriculture, 11*, 159—165.

Liu, Y., Zhao, D. M., Zu, Y. G., Tang, Z. H., Zhang, Z. H., Jiang, Y., et al. (2011). Effects of low light on terpenoid indole alkaloid accumulation and related biosynthetic pathway gene expression in leaves of *Catharanthus roseus* seedlings. *Botanische Studien, 52*, 191—196.

Livadariu, O., Raiciu, D., Maximilian, C., & Căpitanu, E. (2018). Studies regarding treatments of LED-s emitted light on sprouting hemp (*Cannabis sativa* L.). *Romanian Biotechnological Letters*. https://www.rombio.eu/docs/Livadariu%20et%20al.pdf.

Llorente, B., D'Andrea, L., Ruiz-Sola, M. A., Botterweg, E., Pulido, P., Andilla, J., et al. (2015). Tomato fruit carotenoid biosynthesis is adjusted to actual ripening progression by a light-dependent mechanism. *The Plant Journal, 85*(1), 107—119.

Luis, J. C., Perez, R. M., & Gonzalez, F. V. (2007). UV-B radiation effects on foliar concentrations of rosmarinic and carnosic acids in rosemary plants. *Food Chemistry, 101*, 1211—1215.

Magagnini, G., Grassi, G., & Kotiranta, S. (2018). The effect of light spectrum on the morphology and cannabinoid content of *Cannabis sativa* L. *Medical Cannabis and Cannabinoids, 1*, 19—27.

Mahajan, P. V., Caleb, O. J., Gil, M. I., Izumi, H., Colelli, G., Watkins, C. B., et al. (2017). Quality and safety of fresh horticultural commodities: Recent advances and future perspectives. *Food Packaging and Shelf Life, 14*, 2—11.

Mahlberg, P. G., & Hemphill, J. K. (1983). Effect of light quality on cannabinoid content of *Cannabis sativa* L. (Cannabaceae). *Botanical Gazette, 144*, 43—48.

Manaf, H. H., Rabie, K. A. E., & Abd El-Aal, M. S. (2016). Impact of UV-B radiation on some biochemical changes and growth parameters in *Echinacea purpurea* callus and suspension culture. *Annals of Agricultural Science, 61*, 207—216.

Manukyan, A. (2013). Effects of PAR and UV- B radiation on herbal yield, bioactive compounds and their antioxidant capacity of some medicinal plants under controlled environmental conditions. *Photochemistry and Photobiology, 89*, 406—414.

Mariz-Ponte, N., Mendes, R. J., Sario, S., Ferreira de Oliveira, J. M. P., Melo, P., & Santos, C. (2018). Tomato plants use non-enzymatic antioxidant pathways to cope with moderate UV-a/B irradiation: A contribution to the use of UV-a/B in horticulture. *Journal of Plant Physiology, 221*, 32—42.

Massa, G. D., Kim, H. H., Wheeler, R. M., & Mitchell, C. A. (2008). Plant productivity in response to LED lighting. *HortScience, 43*(7), 1951—1956.

Mazza, C. A., & Ballaré, C. L. (2015). Photoreceptors UVR8 and phytochrome B cooperate to optimize plant growth and defense in patchy canopies. *New Phytologist, 207*, 4—9.

Mewis, I., Schreiner, M., Chau Nhi, N., Krumbein, A., Ulrichs, C., Lohse, M., et al. (2012). UV-B irradiation changes specifically the secondary metabolite profile in broccoli sprouts: Induced signaling overlaps with defense response to biotic stressors. *Plant and Cell Physiology, 53*, 1546—1560.

Mitchell, C. A., Dzakovich, M. P., Gomez, C., Lopez, R., Burr, J. F., Hernández, R., et al. (2015). Light-emitting diodes in horticulture. *Horticultural Reviews, 43*, 1−88.

Moon, S. H., Mistry, B., Kim, D. H., & Pandurangan, M. (2017). Antioxidant and anticancer potential of bioactive compounds following UVC light-induced plant cambium meristematic cell cultures. *Industrial Crops and Products, 109*, 762−772.

Moreira-Rodríguez, M., Nair, V., Benavides, J., Cisneros-Zevallos, L., & Jacobo-Velázquez, D. A. (2017). UVA, UVB light, and methyl jasmonate, alone or combined, redirect the biosynthesis of glucosinolates, phenolics, carotenoids, and chlorophylls in broccoli sprouts. *International Journal of Molecular Sciences, 18*(11), 2330−2350.

Mosadegh, H., Trivellini, A., Ferrante, A., Lucchesini, M., Vernieri, P., & Mensuali, A. (2018). Applications of UV-B lighting to enhance phenolic accumulation of sweet basil. *Scientia Horticulturae, 229*, 107−116.

Müller-Xing, R., Xing, Q., & Goodrich, J. (2014). Footprints of the sun: Memory of UV and light stress in plants. *Frontiers of Plant Science, 5*, 1−12.

Muñoz, P., Antón, A., Nuñez, M., Paranjpe, A., Ariño, J., Castells, X., et al. (2007). Comparing the environmental impacts of greenhouse versus open-field tomato production in the Mediterranean region. *Acta Horticulturae, 801*, 1591−1596.

Nascimento, L. B. D. S., Leal-Costa, M. V., Menezes, E. A., Lopes, V. R., Muzitano, M. F., Costa, S. S., et al. (2015). Ultraviolet-B radiation effects on phenolic profile and flavonoid content of *Kalanchoe pinnata*. *Journal of Photochemistry and Photobiology B, 148*, 73−81.

Naznin, M., Lefsrud, M., Gravel, V., & Hao, X. (2016). Different ratios of red and blue LED light effects on coriander productivity and antioxidant properties. *Acta Horticulturae, 1134*, 223−229.

Nelson, J. A., & Bugbee, B. (2013). *Spectral characteristics of lamp types for plant biology.* USA https://www cycloptics.com/sites/default/files/USU_spectral_characteristics pdf.

Nelson, J. A., & Bugbee, B. (2014). Economic analysis of greenhouse lighting: Light emitting diodes vs. High intensity discharge fixtures. *PLoS One, 9*(6), e99010.

Neugart, S., & Schreiner, M. (2018). UVB and UVA as eustressors in horticultural and agricultural crops. *Scientia Horticulturae, 234*, 370−381.

Neugart, S., Zietz, M., Schreiner, M., Rohn, S., Kroh, L. W., & Krumbein, A. (2012). Structurally different flavonol glycosides and hydroxycinnamic acid derivatives respond differently to moderate UV-B radiation exposure. *Physiologia Plantarum, 145*, 582−593.

Nicole, C. C. S., Charalambous, F., Martinakos, S., van de Voort, S., Li, Z., Verhoog, M., et al. (2016). Lettuce growth and quality optimization in a plant factory. *Acta Horticulturae, 1134*, 231−238.

Nishimura, T., Ohyama, K., Goto, E., & Inagaki, N. (2009). Concentrations of perillaldehyde, limonene, and anthocyanin of *Perilla* plants as affected by light quality under controlled environments. *Scientia Horticulturae, 122*, 134−137.

Nishimura, T., Ohyama, K., Goto, E., Inagaki, N., & Morota, T. (2008). Ultraviolet-B radiation suppressed the growth and anthocyanin production of *Perilla* plants grown under controlled environments with artificial light. *Acta Horticulturae, 797*, 425−429.

Olle, M., & Viršile, A. (2013). The effects of light-emitting diode lighting on greenhouse plant growth and quality. *Agricultural and Food Science, 22*(2), 223−234.

Park, S. U., Ahn, D. J., Jeon, H. J., Kwon, T. R., Lim, H. S., Choi, B. S., et al. (2012). Increase in the contents of ginsenosides in raw ginseng roots in response to exposure to 450 and 470 nm light from light-emitting diodes. *Journal of Ginseng Research, 36*, 198−204.

Park, J. S., Choung, M. G., Kim, J. B., Hahn, B. S., Kim, J. B., Bae, S. C., et al. (2007). Gene up-regulated during red colouration in UV-B irradiated lettuce leaves. *Plant Cell Reports, 26,* 507–516.

Paul, N. D., Jacobson, R. J., Taylor, A., Wargent, J. J., & Moore, J. P. (2005). The use of wavelength-selective plastic cladding materials in horticulture: Understanding of crop and fungal responses through the assessment of biological spectral weighting functions. *Photochemistry and Photobiology, 81,* 1052–1060.

Peng, L., Qiang, L., Yunyun, L., Hongjun, Y., & Weijie, J. (2017). Effect of UV-B radiation treatments on growth, physiology and antioxidant systems of cucumber seedlings in artificial climate chamber. *Transactions of the Chinese Society of Agricultural Engineering, 33*(17), 181–186.

Perkins-Veazie, P., Collins, J. K., & Howard, L. (2008). Blueberry fruit response to postharvest application of ultraviolet radiation. *Postharvest Biology and Technology, 47,* 280–285.

Piovene, C., Orsini, F., Bosi, S., Sanoubar, R., Bregola, V., Dinelli, G., et al. (2015). Optimal red: Blue ratio in LED lighting for nutraceutical indoor horticulture. *Scientia Horticulturae, 193,* 202–208.

Piri, E., Babaeian, M., Tavassoli, A., & Esmaeilian, Y. (2011). Effects of UV irradiation on plants. *African Journal of Microbiology Research, 5*(14), 1710–1716.

Prakash, P., & Gupta, N. (2005). Therapeutic uses of *Ocimum sanctum* Linn (tulsi) with a note on eugenol and its pharmacological actions: A short review. *Indian Journal of Physiology & Pharmacology, 49,* 125–131.

Presswood, H. A., Hofmann, R., & Savage, G. P. (2012). Effects of UV-B radiation on oxalate concentration of silver beet leaves. *Journal of Food Research, 1,* 1–6.

Rai, R., Meena, R. P., Smita, S. S., Shukla, A., Rai, S. K., & Pandey-Rai, S. (2011). UV-B and UVC pre-treatments induce physiological changes and artemisinin biosynthesis in *Artemisia annua* L. – an antimalarial plant. *Journal of Photochemistry and Photobiology B: Biology, 105,* 216–225.

Ramani, S., & Chelliah, J. (2007). UV-B-induced signaling events leading to enhanced-production of catharanthine in *Catharanthus roseus* cell suspension cultures. *BMC Plant Biology, 7,* 61–78.

Rechner, O., Neugart, S., Schrener, M., Wu, S., & Poehling, H. M. (2017). Can narrow-bandwidth light from UV-A to green alter secondary plant metabolism and increase *Brassica* plant defenses against aphids? *PLoS One, 12*(11), e0188522.

Rizzini, L., Favory, J.-J., Cloix, C., Faggionato, D., O'Hara, A., Kaiserli, E., et al. (2011). Perception of UV-B by the *arabidopsis* UVR8 protein. *Science, 332*(6025), 103–106.

Robson, T. M., Klem, K., Urban, O., & Jansen, M. A. (2015). Re-interpreting plant morphological responses to UV-B radiation. *Plant, Cell and Environment, 38,* 856–866.

Rodríguez-Calzada, T., Qian, M., Strid, Å., Neugart, S., Schreiner, M., Torres-Pacheco, I., et al. (2019). Effect of UV-B radiation on morphology, phenolic compound production, gene expression, and subsequent drought stress responses in chili pepper (*Capsicum annuum* L.). *Plant Physiology and Biochemistry, 134,* 94–102.

Rouphael, Y., Kyriacou, M. C., Petropoulos, S. A., De Pascale, S., & Colla, G. (2018). Improving vegetable quality in controlled environments. *Scientia Horticulturae, 234,* 275–289.

Sabzalian, M. R., Heydarizadeh, P., Zahedi, M., Boroomand, A., Agharokh, M., Sahba, M. R., et al. (2014). High performance of vegetables, flowers, and medicinal plants in a red-blue

LED incubator for indoor plant production. *Agronomy for Sustainable Development, 34,* 879−886.

Sakalauskaitė, J., Viškelis, P., Dambrauskienė, E., Sakalauskienė, S., Samuolienė, G., Brazaitytė, A., et al. (2013). The effects of different UV-B radiation intensities on morphological and biochemical characteristics in *Ocimum basilicum* L. *Journal of the Science of Food and Agriculture, 93,* 1266−1271.

Sakalauskaitė, J., Viškelis, P., Duchovskis, P., Dambrauskienė, E., Sakalauskienė, S., Samuolienė, G., et al. (2012). Supplementary UV-B irradiation effects on basil (*Ocimum basilicum* L.) growth and phytochemical properties. *Journal of Food Agriculture and Environment, 10,* 342−346.

Samuolienė, G., Brazaitytė, A., & Vaštakaitė, V. (2017). Light-emitting diodes (LEDs for improved nutritional quality. In S. D. Gupta (Ed.), *Light emitting diodes for agriculture: Smart lighting* (p. 334). Springer Nature Singapore Ltd. https://doi.org/10.1007/978-981-10-5807-3.

Samuolienė, G., Urbonavičiūtė, A., Brazaitytė, A., Šabajevienė, G., Sakalauskaitė, J., & Duchovskis, P. (2011). The impact of LED illumination on antioxidant properties of sprouted seeds. *Central European Journal of Biology, 6,* 68−74.

Schreiner, M., Korn, M., Stenger, M., Holzgreve, L., & Altmann, M. (2013). Current understanding and use of quality characteristics of horticulture products. *Scientia Horticulturae, 163,* 63−69.

Schreiner, M., Mewis, I., Huyskens-Keil, S., Jansen, M. A. K., Zrenner, R., Winkler, J. B., et al. (2012). UV-B-induced secondary plant metabolites-potential benefits for plant and human health. *Critical Reviews in Plant Sciences, 31,* 229−240.

Shiga, T., Shoji, K., Shimada, H., Hashida, S., Goto, F., & Yoshihara, T. (2009). Effect of light quality on rosmarinic acid content and antioxidant activity of sweet basil, *Ocimum basilicum* L. *Plant Biotechnology, 26,* 255−259.

Suchar, V. A., & Robberecht, R. (2015). Integration and scaling of UV-B radiation effects on plants: From DNA to leaf. *Ecology and Evolution, 5,* 2544−2555.

Sun, R., Hikosaka, S., Goto, E., Sawada, H., Saito, T., Kudo, T., et al. (2012). Effects of UV irradiation on plant growth and concentrations of four medicinal ingredients in Chinese licorice (*Glycyrrhiza uralensis*). *Acta Horticulturae, 956,* 643−648.

Takshak, S., & Agrawal, S. B. (2014). Secondary metabolites and phenylpropanoid pathway enzymes as influenced under supplemental ultraviolet-B radiation in *Withania somnifera* Dunal, an indigenous medicinal plant. *Journal of Photochemistry and Photobiology B, 140,* 332−343.

Takshak, S., & Agrawal, S. B. (2015). Defence strategies adopted by the medicinal plant *Coleus forskohlii* against supplemental ultraviolet-B radiation: Augmentation of secondary metabolites and antioxidants. *Plant Physiology and Biochemistry (Paris), 97,* 124−138.

Taulavuori, K., Pyysalo, A., Taulavuori, E., & Julkunen-Tiitto, R. (2018). Responses of phenolic acid and flavonoid synthesis to blue and blue-violet light depends on plant species. *Environmental and Experimental Botany, 150,* 183−187.

Tegelberg, R., Julkunen-Tiitto, R., & Aphalo, P. (2004). Red: Far-red light ratio and UV-B radiation: Their effects on leaf phenolics and growth of silver birch seedlings. *Plant, Cell and Environment, 27,* 1005−1013.

Teramura, A. H., Tevini, M., & Iwanzik, W. (1983). Effects of ultraviolet-B irradiation on plants during mild water stress. *Physiologia Plantarum, 57,* 175−180.

Tossi, V., Amenta, M., Lamattina, L., & Cassia, R. (2011). Nitric oxide enhances plant ultra-violet - B protection up-regulating gene expression of the phenylpropanoid biosynthetic pathway. *Plant, Cell and Environment, 34*, 909—921.

Tsormpatsidis, E., Henbest, R. G. C., Battey, N. H., & Hadley, P. (2010). The influence of ul-traviolet radiation on growth, photosynthesis and phenolic levels of green and red lettuce: Potential for exploiting effects of ultraviolet radiation in a production system. *Annals of Applied Biology, 156*(3), 357—366.

Tsormpatsidis, E., Henbest, R. G. C., Davis, F. J., Battey, N. H., Hadley, P., & Wagstaffe, A. (2008). UV irradiance as a major influence on growth, development and secondary prod-ucts of commercial importance in Lollo Rosso lettuce 'Revolution' grown under polyeth-ylene films. *Environmental and Experimental Botany, 63*, 232—239.

Urban, L., Chabane Sari, D., Orsal, B., Lopes, M., Miranda, R., & Aarrouf, J. (2018). UV-C light and pulsed light as alternatives to chemical and biological elicitors for stimulating plant natural defenses against fungal diseases. *Scientia Horticulturae, 235*, 452—459.

Urban, L., Charles, F., de Miranda, M. R. A., & Aarrouf, J. (2016). Understanding the phys-iological effects of UV-C light and exploiting its agronomic potential before and after harvest. *Plant Physiology and Biochemistry, 105*, 1—11.

Vaštakaitė, V., Viršilė, A., Brazaitytė, A., Samuolienė, G., Jankauskienė, J., Sirtautas, R., et al. (2015). The effect of UV-A supplemental lighting on antioxidant properties of *Ocimum basilicum* L. microgreens in greenhouse. In A. Raupalienė (Ed.), *Proceedings of the 7th international scientific conference rural development 2015*. https://doi.org/10.15544/RD.2015.031.

Verdaguer, D., Jansen, M. A., Morales, L. O., & Neugart, S. (2017). UV-A radiation effects on higher plants: Exploring the known unknown. *Plant Science, 255*, 72—81.

Viršilė, A., Olle, M., & Duchovskis, P. (2017). LED lighting in horticulture. In S. D. Gupta (Ed.), *Light emitting diodes for agriculture: Smart lighting* (p. 334). Springer Nature Singapore Ltd. https://doi.org/10.1007/978-981-10-5807-3.

Wargent, J. J. (2016). UV LEDs in horticulture: From biology to application. *Acta Horticulturae, 1134*, 25—32.

Wargent, J. J., Elfadly, E. M., Moore, J. P., & Paul, N. D. (2011). Increased exposure to UV-B radiation during early development leads to enhanced photoprotection and improved long-term performance in *Lactuca sativa*. *Plant, Cell and Environment, 34*, 1401—1413.

Wargent, J. J., Moore, J. P., Roland Ennos, A., & Paul, N. D. (2009). Ultraviolet radiation as a limiting factor in leaf expansion and development. *Photochemistry and Photobiology, 85*(1), 279—286.

Wargent, J., Nelson, B., Mcghie, T., & Barnes, P. (2015). Acclimation to UVB radiation and visible light in *Lactuca sativa* involves up-regulation of photosynthetic performance and orchestration of metabolome-wide responses. *Plant, Cell and Environment, 38*, 929—940.

Wu, D., Hu, Q., Yan, Z., Chen, W., Yan, C., Huang, X., et al. (2012). Structural basis of ultraviolet-B perception by UVR8. *Nature, 484*, 214—219.

Yang, X., Xu, H., Shao, L., Li, T., Wang, Y., & Wang, R. (2018). Response of photosynthetic capacity of tomato leaves to different LED light wavelength. *Environmental and Experi-mental Botany, 150*, 161—171.

Yin, R., & Ulm, R. (2017). How plants cope with UV-B: From perception to response. *Current Opinion in Plant Biology, 37*, 42—48.

Zobayed, S. M., Afreen, F., & Kozai, T. (2005). Necessity and production of medicinal plants under controlled environments. *Environment Control in Biology, 43*(4), 243—252.

UV LEDs in Postharvest Preservation and Produce Storage

4

**Aušra Brazaitytė, Viktorija Vaštakaitė-Kairienė, Neringa Rasiukevičiūtė,
Alma Valiuškaitė**

Institute of Horticulture, Lithuanian Research Centre for Agriculture and Forestry, Lithuania

Chapter outline

Introduction 68
UV for Higher Nutritional Value and Improved Postharvest Quality 68
    Chlorophylls 69
    Phenolic Compounds and Antioxidant Activity 70
    Enzymes 72
The Storage of Horticulture Production under Artificial Light Sources 73
    The Conventional Lamps Versus LEDs 73
        Color Attributes 73
        Texture 74
        Biochemical Attributes 75
        Enzymes 75
        Vitamins 76
        Phenolic Compounds and Antioxidant Activity 76
        Carotenoids 77
    Fungal Control 78
Conclusions 83
Acknowledgments 84
References 84

Abstract

The application of ultraviolet (UV) irradiation at pre- and postharvest stages for accumulation of health-beneficial properties having biologically active substances and extension of shelf life as well as prevention against fungal diseases of horticultural products is overviewed. The studies conducted on application of UV light-emitting diode (LED) lighting technology to improve and preserve quality of vegetables and fruits during storage are discussed. To confirm the need for further studies on UV LED lighting for storage quality, the research-based knowledge about non-LED artificial UV light sources was reviewed.

Ultraviolet LED Technology for Food Applications. https://doi.org/10.1016/B978-0-12-817794-5.00004-2
Copyright © 2019 Elsevier Inc. All rights reserved.

Keywords: Fruits; Fungi; Pathogens; Postharvest; Preharvest; Storage quality; UV; UV LED; Vegetables.

Introduction

Fruits and vegetables containing health-beneficial bioactive compounds such as phenolic acids, flavonoids, anthocyanins, glucosinolates, tocopherols, ascorbate, carotenoids, and others are a necessary part of the human diet. Fact is that nutritional value of fresh horticultural products is determined at harvest through various environmental and agronomic factors that are essential during their cultivation with the purpose to enhance phytochemicals content and improve their composition (Charles & Arul, 2007; Hewet, 2006). On the other hand, fruits and vegetables are very perishable and extending of shelf life, maintaining resistance to diseases, and the preservation of beneficial phytochemicals, is a significant concern during the postharvest stage. Various technologies of shelf life extension are applied for maintenance of horticulture production, including physical, chemical, and biologic preservation methods (Ma, Zhang, Bhandari, & Gao, 2017; Rouphaela, Kyriacou, Petropoulos, Pascale, & Coll, 2018). The ultraviolet (UV) irradiation is one of the most promising alternatives to chemical approaches and has great potential for extension of shelf life of various horticultural plants during storage. Low doses of UV—from 0.125 to 9 kJ m^{-2}—are beneficial for disease resistance, delay ripening and senescence, improve phytochemicals content, and are termed as hormetic doses (Charles & Arul, 2007; Fonseca & Rushing, 2008; Urban, Charles, de Miranda, & Aarrouf, 2016). Many studies about the effect of UV irradiation are concerned with short-term UVC exposure at the postharvest stage. On the other hand, although it is the well-known impact of UVA and UVB irradiation on plants including horticultural, it is lack of information about studies of these wavelengths for postharvest applications (Charles & Arul, 2007; Fonseca & Rushing, 2008; Ma et al., 2017). It should be noted that for most UV treatments of the fruits and vegetables at postharvest stage, space-consuming UV-emitting features were used with a wide range of waveband. Advantages of UV light-emitting diode (LED) light such as energy efficient, small size, no damage for human eyes and skin, convenient manipulation of UV spectrum, and no production of mercury waste increased interest to apply them during storage. However, an overview of the literature showed that data about UV LED application concerning horticultural research and production at postharvest are still at the initial stage (D'Souza, Yuk, Khoo, & Zhou, 2015; Mitchell et al., 2015; Wargent, 2016).

UV for Higher Nutritional Value and Improved Postharvest Quality

Horticulture plants are a source of nutritionally important phytochemicals, such as vitamins, minerals, secondary metabolites, which are beneficial for human health.

Various preharvest factors, including genetic material, environmental factors, and crop management technologies, are used for increasing accumulation of bioactive compounds in plants (Bian, Yang, & Liu, 2015; Hewett, 2006; Ilić & Fallik, 2017; Rouphael et al., 2018). Light is a key factor, which influences the external and internal quality of various horticulture plants. Investigations about manipulation of light quantity and quality using photo-selective netting or films, high-intensity discharge (HID) lighting, and LED light technology for improving the yield, quality, and phytochemical composition of cultivated plants are described in scientific literature, discussed in some of review articles (Bantis et al., 2018; Bian, Yang, & Liu, 2015; Carvalho & Folta, 2014; Demotes-Mainard et al., 2016; Ilić & Fallik, 2017; Huché-Thélier et al., 2016; Mitchell et al., 2015; Olle & Viršile, 2013; Wargent, 2016) and in Chapter 3 of this book. However, there is a lack of information how vegetable quality determined by light conditions changes during storage when postharvest stresses such as the wounding due to cutting and senescence due to dark storage are occurred (Witkowska, 2013).

Although UV LED light is still a new technology, a few publications exist about uses of only UVA LED light as a part of different lighting systems or as a sole source of light and their effect on content of bioactive compounds in various leafy vegetables (Brazaityte et al., 2016; Brazaitytė et al., 2015; Goto, Hayashi, Furuyama, Hikosaka, & Ishigami, 2016; Lee et al., 2010; Li and Kubota; 2009; Samuoliene et al., 2013; Vaštakaitė et al., 2016). No data were found about the effect of near to far UV light LED wavelength irradiation at preharvest or harvest stage on shelf life of vegetables during storage in dark. Literature data showed that mostly UVB and UVC lamps, such as mercury, xenon, and others, are used for short-term exposure at such stages with the purpose to prolong shelf life and internal quality of various vegetables at the postharvest stage.

Chlorophylls

One of the essential indexes of vegetable quality during storage is degradation of chlorophyll, which occurs during leaf senescence, fruit ripening, and during biotic and abiotic stresses. During degradation process, chlorophyll a is transformed in chlorophyllide a by the action of the enzyme chlorophyllase, and chlorophyllide a is transformed into pheophorbide a by Mg-dechelatase. Tetrapyrrolic rings are broken producing noncolored chlorophyll derivatives (Hörtensteiner & Kräutler, 2011; Matile, Hörtensteiner, & Thomas, 1999). Otherwise, such degradation could be caused by peroxidases, which produce other chlorophyll derivatives in an oxidative reaction (Funamoto, Yamauchi, Shigenaga, & Shigyo, 2002). Loss of chlorophyll in pak choi was low ($<1.8 \, g \, cm^{-2}$, $<16\%$) and UVB irradiated, and nonirradiated plants showed the same decreasing trend during storage at $2°C$ under a controlled and normal air atmosphere (Harbaum-Piayda et al., 2010). Aiamla-or, Kaewsuksaeng, Shigyo, and Yamauchi (2010) reported that short-term UVB irradiation of 8.8 kJ m^{-2} at harvest stage efficiently delayed the decrease of the contents of chlorophylls *a* and *b* in broccoli during storage at 15°C. Topcu et al. (2015) stated that photosynthetic pigment contents of broccoli decreased with increased doses of UVB from 2.2 till 16.4 kJ m^{-2}. However, chlorophyll *a* and carotenoids content of such plants decreased further during storage at 0°C, but chlorophyll *b* content is increased.

Overview of literature sources showed that UVC (200−280 nm) treatments, contrary to UVB (280−320 nm), are mostly related to short-term exposure at harvest stage of various vegetables and after-effect during storage. According to literature, short-term UVC irradiation at harvest stage prevents chlorophyll degradation in broccoli (at 20°C, UVC dose of 10 kJ m^{-2}) (Costa, Vicente, Civello, Chaves, & Martínez, 2006), garden cress (UVC for 30 min) (Kasim & Kasim, 2012), Chinese kale (at 20°C, UVC dose of 3.6 and 5.4 kJ m^{-2}) (Chairat, Nutthachai, & Varit, 2013). However, chlorophyll content maintained at the same level during shelf life in tatsoi baby leaves (Tomás-Callejas, Otón, Artés, & Artés-Hernández, 2012) and minimally processed spinach (Artés-Hernández, Escalona, Robles, Martínez-Hernández, & Artés, 2009). Meanwhile, UVC irradiation delayed not only chlorophyll degradation but also the increase of chlorophyllase and chlorophyll-peroxidase activity in broccoli (Costa et al., 2006). Chairat et al. reported that chlorophyllase, Mg-dechelatase, and chlorophyll-degrading peroxidase activities increased in Chinese kale (Chairat et al., 2013).

Phenolic Compounds and Antioxidant Activity

The UVB light is not the only part of total solar radiation, which could cause significant biologic damage in plants but could induce specific plant responses, some of which are desirable for horticultural purposes. One positive effect of UVB irradiation is the stimulation of secondary metabolism, including the production of beneficial health phytochemicals, such as phenolics, flavonoids, anthocyanins, etc. However, such irradiation is most effective, when exposure lasted during more extended period, usually several hours or days. It is often problematic to use it practically for preharvest treatment of various vegetables with purpose to extend their shelf life and as evidenced low amount of reports published on this topic (Huché-Thélier et al., 2016; Jansen, Hectors, O'Brien, Guisez, & Potters, 2008; Topcu et al., 2015). Harbaum-Piayda et al. (2010) reported positive the 10 days effect of UVB irradiation of 0.35−0.42 W m^{-2} during the growth period on polyphenolic content in pak choi. It is known that various polyphenols have protective functions in plants and their content is closely related to several stress factors such as UVB irradiation and low temperature. Meanwhile, these metabolites also are essential for crop quality and beneficial for human health (Harbaum-Piayda et al., 2010; Topcu et al., 2015). The content of polyphenols and flavonoids continuously increased during cold storage (2°C) at both controlled and normal air atmosphere (Harbaum-Piayda et al., 2010). Quite contrary results were determined for broccoli storage at a temperature of 0°C, when exposure of 2.2, 8.8, and

16.4 kJ m^{-2} of UVB irradiation was used during growth. Although the total phenolic and flavonoid content increased after UV exposure, it decreased during storage (Topcu et al., 2015).

Among all phenolic compounds such as simple phenols, flavonoids, phenolic acids, lignins, and tannins, UVC exposure mostly stimulate synthesis of flavonoids (Urban et al., 2016). UVC (10 kJ m^{-2})-treated broccoli had higher antioxidant capacity and content of total phenols and flavonoids during storage (Costa et al., 2006). Such irradiance did not affect the phenolic content and the total antioxidant capacity in tatsoi baby leaf, which were kept at the same level during storage (Tomás-Callejas et al., 2012). According to Artes-Hernandez et al. (2009), UVC irradiation of 0, 4.54, 7.94, and 11.35 kJ m^{-2} of minimally processed spinach prior packaging resulted in a decrease of total antioxidant capacity content during shelf life being more obvious when leaves were stored at 8°C rather than at 5°C. The total polyphenol content decreased steadily during the storage period and was more evident at high UVC doses (11.35 kJ m^{-2}) and temperature (8°C) when their content decreased from 2.6 till 1.6 (g kg^{-1} FW).

Anthocyanins are important polyphenols due to their anti-inflammatory activity. Purple-red colors of fruits and vegetables resulted in the accumulation of abundant cyanidin-based anthocyanins (Jagadeesh et al., 2011; Wiczkowski, Szawara-Nowak, & Topolska, 2013). Wu et al. (2017) determined 15 cyaniding derivatives in UVC-treated fresh-cut red cabbage, and four of these were anthocyanins absent in control samples. The optimum dose of UVC for increase of total anthocyanin content in fresh-cut red cabbage was 3.0 kJ m^{-2}, and this effect remained during the prolonged storage period when their content peak at 8 days of storage was 105.42 ± 3.63 mg g^{-1} DW. Different UVC irradiation doses (0, 1.0, 3.0, or 5.0 kJ m^{-2}) resulted in various responses for each anthocyanin compounds, and these changes were dose dependent (Wu et al., 2017).

According to literature sources, UVC irradiation also affects the content of various phenolic compounds in tomato, pepper, and strawberry. Jagadeesh et al. (2011) reported that total phenolic contents (TPCs) were higher in the UVC (3.7 kJ m^2)-treated tomatoes, but such dose did not affect the antioxidant activity significantly. Other data showed that the content of bioactive compounds in tomato fruits was UVC dose dependent. Liu, Cai, Lu, Han and Ying (2012) determined that UVC irradiation at 4 or 8 kJ m^{-2} stimulated the accumulation of total flavonoids and increased the antioxidant activity. Athough 2 or 16 kJ m^{-2} UVC irradiation also enhanced antioxidant activity, but at a lesser level. These authors identified seven phenolic compounds (gallic acid, (+)-catechin, chlorogenic acid, caffeic acid, syringic acid, p-coumaric acid, and quercetin) in tomato fruit, and most of the UVC irradiation of 4 or 8 kJ m^{-2} significantly increased. Pérez-Ambrocio et al. (2018) stated that UVC light might stimulate the synthesis of total flavonoids, phenolic compounds, capsaicin, and the antioxidant capacity in habanero pepper. According to Erkan, Wang, and Wang (2008), all UVC doses (0.43, 2.15, and 4.30 kJ m^{-2}) increased the content of phenolic and total anthocyanin in strawberry fruits.

Enzymes

It is known that various environmental stress, including UVC, cause the production and accumulation of reactive oxygen species (ROS) in many plant species. The main ROS—singlet oxygen, hydrogen peroxide, and hydroxyl radicals—are produced because of biochemical and physicochemical reactions in plant cells and are one of the important factors involved in the senescence process. To protect plants from the harmful effect of toxic concentrations of ROS, they have an efficient antioxidative defense system, composed of nonenzymatic and enzymatic components in all plant cells (Barka, 2001; Urban et al., 2016). The decrease in activities of antioxidant enzymes, particularly peroxidase and superoxide dismutase, was delayed under the effect of UVC dose 3.6 and 5.4 kJ m^{-2} in Chinese kale (Chairat et al., 2013). Barka (2001) investigated the changes in the activities of different enzymes, such as guaiacol and ascorbate peroxidases, catalase, superoxide dismutase, ascorbate oxidase, lipoxygenase, and phenylalanine ammonia lyase in tomato fruit affected by 3.7 kJ m^{-2} of UVC. Obtained results showed that such irradiation caused an increase of the guaiacol peroxidase and ascorbate peroxidase activities, but catalase activity was like the control.

Meanwhile, the activities of superoxide dismutase and ascorbate oxidase were reduced after UVC exposure and activities of lipoxygenase and phenylalanine ammonia lyase increased within the first 5 days, followed by the second period where these activities were below those of the control. It was suggested that UVC induced accumulation of photooxidation products, which activated defense mechanisms against oxidation and resulted in the delay of ripening and senescence of irradiated tomato fruit (Barka, 2001). Erkan et al. (2008) reported that UVC irradiation of 2.15 and 4.30 kJ m^{-2} enhanced the activities of all determined antioxidant enzymes and nonenzymic components in strawberry.

Preharvest UV irradiation affected the changes of other phytochemicals in broccoli during storage. Ascorbic acid (AA) increased with increasing UVB dose from 0 till 16.4 kJ m^{-2}, but prolonged storage till 60 days negatively affected its content (Topcu et al., 2015). The content of glucosinolates, which according to literature data are linked to reduced incidence of some cancers (Bjorkman et al., 2011), showed an increase till the 45th day of storage, and later their levels began to decline (Topcu et al., 2015). The same authors reported that total soluble solids, solids content, and titratable acidity (TA) decreased during storage till 60 days and were significantly negatively affected by increasing (from 0 till 16.4 kJ m^{-2}) UVB doses, except TA. Jagadeesh et al. (2011) reported that AA content was higher in the UVC (3.7 kJ m^2)-treated tomatoes, but such dose significantly reduced the lycopene content.

In summary, application of UV irradiation at preharvest stage is useful for decreasing senescence and increasing accumulation of health-beneficial bioactive compounds. However, UV response is related to interactions between various environmental factors such as dose, exposure time, storage temperature, cultivar, etc. As the overview literature sources showed, to research in this area, there is not much and further studies in different conditions should be performed.

The Storage of Horticulture Production under Artificial Light Sources

The Conventional Lamps Versus LEDs

After harvest, a reduction in the sources of energy, water, and nutrients leads to rapid initiation of fruit and vegetable senescence (King & Morris, 1994). The main senescence symptoms of fruits and vegetables are discoloration, firmness, and shriveling due to water loss. Ultraviolet lighting treatment could be effective on maintaining attractive fruits and vegetables; however, the duration of UV light exposure and postharvest storage at certain temperature must be considered.

Color Attributes

There are not many studies that have investigated the effect of near ultraviolet (UVA) LED lighting in the postharvest storage of fruits and vegetables. The effectiveness of unfiltered germicidal emitting UVA lamps to maintain color attributes of cucumber is discussed (Kasim & Kasim, 2008). The L* values of cucumber cv. "Silor" treated with UVA light for 30 min were higher in nontreated samples or treated with 5-, 10-, or 20-min exposure of UVA light. However, L* value was the highest of nontreated cucumbers at the end of storage. The same tendency was observed for a* value of cucumber treated with UVA light exposure for 30 min (at 5°C) compared with nontreated and treated for a shorter period of UVA exposure. At the 10°C, however, a* values of cucumbers treated with UVA for 30 min was lower than those of the other samples, and differences were significant at the end of the storage. The changes in b* values of nontreated and treated with UVA light cucumbers were similar to the a* values of samples. The highest Hue angle was reported for cucumbers treated with UVA irradiation for 30 min, and the lowest of treated UVA for 5 min.

The effects of UVA LED light on the postharvest quality of fruits and vegetables could be similar to the blue LED light (405—500 nm) because of the common photosensory system of photoreceptors cryptochromes (Lin and Todo, 2015). The main parameters studied and reported were changes in color of produce expressed in the scale of L (lightness), a (redness), and b (yellow). For example, blue LED light decreased fruit lightness (L*), green color a* [(+) redness/(−) greenness], and increased b* [(+) yellowness/(−) blueness] value in cabbage cv. "Dongdori" (Lee, Ha, Oh, & Cho, 2014). The color changes in strawberry cv. "Fengguang" expressed as color index red grapes (CIRG) value (180—h/C + L) tended to increase gradually in blue light—treated fruit at 470 nm (40 μmol m^{-2} s^{-1}) compared with control (Xu et al., 2014). The white and blue LED light (20 μmol m^{-2} s^{-1}) illuminated broccoli cv. "Legacy" demonstrated lower L* value, chlorophyll *a* and *b* content compared with control, while Hue angle was higher in treated samples (Hasperué, Guardianelli, Rodoni, Chaves, & Martínez, 2016). The same effect on lower L* value and Hue angle was observed in Brussel sprouts treated with blue and white LED light. The decrease of L* value resulted in delayed yellowing of sprouts (Hasperué, Rodonia, Guardianelli, Chaves, & Martínez, 2016). Mature green tomato cv.

"Dotaerang" irradiated with blue 440–450 nm LED light (85.7 µmol m^{-2} s^{-1}) demonstrated less increased a* and high increased b* values indicating that the blue light delayed the maturation process compared with the fruit stored in the dark. This observation means that blue LED light delayed the maturation process of green tomato (Dhakal, Baek, 2014). In contrast, blueberries illuminated with blue 480 nm LED light (27 µW cm^{-2}) had no exceptional effect on L* value compared with darkness (Spotts, Sandhu, Singh, & Collier, 2017). In addition, blue 470 nm LED light (50 µmol m^{-2} s^{-1}) did not affect the yellowing process of broccoli florets compared with white light; however, the florets displayed obvious yellowing in the control and after blue light LED treatment (Ma et al., 2014).

The a* value (redness) of UVC-treated tomatoes cv. "Red ruby" using germicidal lamps (254 nm, 13.7 kJ m^{-2}) on breaker stage (mature green color) was significantly lower throughout the storage compared with untreated samples. There was no significant effect of light on a b* value between untreated and treated tomatoes at the end of storage. The Hunter a*/b* ratio of surface color is considered as a reference parameter for red color development in tomatoes (Arias, Lee, Logendra, & Janes, 2000). The a*/b* ratios of UVC light–treated samples were slightly lower than those of the untreated samples (Liu, Zabaras, Bennett, Aguas, & Woonton, 2009). The UVC light emitted by the germicidal lamp (254 nm) led to lower a* and higher °h values of tomato cv. "Zinac" at green maturity stage (Pinheiro, Alegria, Abreu, Gonçalves, & Silva, 2015).

Texture

The tissue turgidity, which is mainly determined by the degree of dehydration, is linked to postharvest quality during the storage (Serrano, Martinez-Romero, Guillen, Castillo, & Valero, 2006). Flesh firmness is a component of texture, which is a complex sensory attribute that also includes crispiness and juiciness (Konopacka & Plocharski, 2003). The UVA treatment was effective to maintain the flesh firmness of cucumber cv. "Silor," and there were no significant differences on weight loss (WL) for nontreated, and UVA-treated cucumbers (Kasim & Kasim, 2008). The effects of blue light on texture and WL attributes were studied in tomatoes, broccoli, cucumbers, and peppers. Blue 440–450 nm LED light (85.7 µmol m^{-2} s^{-1}) had negative effect by decreasing the firmness of tomato cv. "Dotaerang" fruit after 7 days of treatment (Dhakal & Baek, 2014). The higher WL was in broccoli cv. "Legacy" stored under continuous white and blue LED illumination (20 µmol m^{-2} s^{-1}) compared with nontreated group (Hasperué, Guardianelli et al., 2016). Brussels sprouts exposed to continuous white and blue LED light showed higher WL compared with controls stored in dark until 10 days of storage (Hasperué, Rodonia et al., 2016). The greater WL in light-exposed samples could be due to possible higher stomata openings caused by blue light irradiation (Kinoshita et al., 2001; Noichinda, Bodhipadma, Mahamontri, Narongruk, & Ketsa, 2007). In addition, bell pepper cv. "Jubilee F1" irradiated with 2.4 kJ m^{-2} UVC (254 nm) had the highest amount of WL, and the lowest WL was obtained from 3.6 kJ m^{-2} UVC-treated fruit (Karaşahin Yildirim & Pekmezci, 2017).

Biochemical Attributes

Plants use sugars as the primary source of energy to perform numerous chemical reactions necessary to the maintenance of tissue integrity and the synthesis of new compounds (De Vries, 1975). Strawberries cv. "Suljyang" treated with UVA LED light (at 385 nm) showed increased total soluble solids (TSS) compared with control or blue, green, and red LED light (Kim et al., 2011). The content of TSS was higher in UVA-treated cucumber compared with nontreated at 10°C after 15 days (Kasim, Kasim, 2008). Panjai, Noga, Fiebig, and Hunsche (2017) evaluated the effects of short periods of daily UV radiation on the postharvest quality of green tomatoes cv. "Cappricia." The fruit was irradiated with LED modules consisted of the 60% UVB (290 nm), 30% UVA (320−400 nm), 4% UVC (200−280), and 6% visible light. Tomato fruits were additionally irradiated with UV tubes' light for 15 min twice per day (5.53 kJ m^{-1}), and the red 665 nm light (113 μmol m^{-2}) was applied for the whole storage period. The significantly higher concentration of TSS was in tomato fruit exposed to red light and red light with UV for 5 days compared with darkness and those treated with UV only. However, there was no significant difference in TSS between the treatments at 20 days after harvest.

The broccoli cv. "Legacy" irradiation with white and blue LED light suppressed a decrease of total sugar content and showed higher amounts of glucose and fructose after 2 and 3 days, and higher amounts of sucrose after 4 days of storage at 22°C. During the storage period at 5°C, broccoli heads stored under illumination of white and blue light maintained the fructose and glucose levels and significantly increased a level of sucrose. The higher sugar accumulation in broccoli head stored under white and blue LED light at 5°C could be because of lower energy needed for the maintenance of metabolism in a colder condition that under 22°C (Hasperué, Guardianelli et al., 2016). The outer leaves of Brussels sprouts exposed to white and blue LED light demonstrated higher TSS than control (dark) until day 5 at 22°C (Hasperué, Rodonia et al., 2016).

The decrease of TSS content was found in bell pepper "Jubilee F1" treated with 3.6 kJ m^{-2} UVC dosages compared with lower dosage (1.3 and 2.4 kJ m^{-2}) of UVC (Karaşahin Yildirim & Pekmezci, 2017).

TA in both control and treated with blue 470 nm (40 μmol m^{-2} s^{-1}, 5°C, 12 days) in strawberry cv. "Fengguang" fruit continuously decreased as the storage time progressed. However, the higher TA was in fruit treated with blue light (Xu et al., 2014). The UVC dosage 3.6 kJ m^{-2} resulted in significantly lower values of TA in bell pepper, and the highest values detected with 1.3 and 2.4 kJ m^{-2} UVC dosages during storage period (Karaşahin Yildirim & Pekmezci, 2017).

Enzymes

The superoxide dismutase, catalase, and ascorbate peroxidase activities in blue 470-nm light-treated (40 μmol m^{-2} s^{-1}) strawberry cv. "Fengguang" fruit was significantly higher than in the control fruit during the storage period. Significantly lower levels of the ROS such as O^{2-} production and H_2O_2 content were observed in blue 470-nm light irradiated fruit compared with control after 4 days of storage. In addition, blue 470-nm light treatment delayed the increase of maleic dialdehyde level in strawberry, and the content was always lower than that in control (Xu et al., 2014).

Vitamins

The significantly higher vitamin C content was observed in cabbage cv. "Dongdori" after irradiation with blue LED compared with non-LED control (Lee et al., 2014). However, the content of AA in strawberry cv. "Fengguang" fruit diminished both in blue 470-nm light-treated (40 µmol m^{-2} s^{-1}) and control throughout the storage period. A higher content of AA was observed after 4 days of storage of treated fruits (Xu et al., 2014). In addition, the slightly higher content of AA in broccoli cv. "Legacy" was observed due to the irradiation with white and blue LED light (Hasperué, Guardianelli et al., 2016). The blue 470-nm LED light (50 µmol m^{-2} s^{-1}) did not significantly affect the content of L-dehydroascorbic acid (DHA) in broccoli (Ma et al., 2014). The treatment using combined cool white light (19 µmol m^{-2} s^{-1}) and UVB (19–22 kJ m^{-2} d^{-1}) did not affect vitamin C, AA, and DHA contents in broccoli cv. "Marathon" flower buds during storage (Rybarczyk-Plonska et al., 2014).

Phenolic Compounds and Antioxidant Activity

The beneficial effect of light was reported on secondary metabolites such as phenolic compounds that play a beneficial role in human health promotion. The additional exposure to UV light led to significantly decreased flavonoid concentration in tomato fruit 5 days after harvest, and the concentration was the lowest for all days until the end of the sampling period. In addition, there was no significant difference in flavonoid concentration between additional UV, red light, and red light with UV exposure 20 days after harvest. In addition, fruit treated with additional UV light had significantly lower content of phenols at 5 days after harvest and slowly recovered to concentrations evaluated directly after harvest until 20 days (Panjai et al., 2017).

Exposure to blue LED light increased the TPC in cabbage cv. "Dongdori" in comparison with non-LED (Lee et al., 2014). The TPC in strawberry cv. "Fengguang" fruit with or without blue 470 nm (40 µmol m^{-2} s^{-1}) treatment increased gradually with storage time. Blue 470 nm light treatment maintained significantly higher TPC after 2 days of storage (Xu et al., 2014). In contrary, blue 480 nm LED light (27 µW cm^{-2}) decreased TPC with the increase in treatment time for 2, 4, and 6 h in blueberries (Spotts et al., 2017). The exposure with white and blue LED light led to the highest contents of flavonoids in outer and inner leaves of Brussels sprouts. The higher content of flavonoids could be a result of the especially enriched spectrum with blue light (Hasperué, Rodonia et al., 2016).

The TPC content in tomato cv. "Zinac" increased after the treatments with UVC at 0.32, 0.97, 2.56, 4.16, and 4.83 kJ m^{-2} doses compared with content determined at the ninth day and first day of storage. The TPC of UVC-treated tomato at 0.32 kJ m^{-2} was 55% higher than that observed on control samples. However, by the last day of storage, the TPC of all UVC samples reached a similar value to control samples. UVC treatment increased the TPC through the adaptation mechanism of tomato due to UV stress, which promoted the enzymatic activity of phenylalanine ammonia-lyase, a key enzyme in the production of phenylpropanoids, leading to an increase of phenols. However, the reduction of the TPC can be the result of UVC radiation action on phenylpropanoid pathway responsible for phenolics synthesis, or due to polyphenoloxidase enzymatic activity that oxidizes phenolics into quinones polymers (Pinheiro et al., 2015).

The antiradical scavenging activity measured as DPHH (2,2-diphenyl-1-picrylhydrazyl) usually correlates with TPC. In strawberry cv. "Fengguang" fruit treated with blue light at 470 nm (40 μmol m^{-2} s^{-1}), DPPH radical scavenging activity increased gradually with the storage time. The significantly higher DPPH radical scavenging activity was determined in a fruit treated with blue 470 nm light than in untreated control after 4 days storage (Xu et al., 2014). In broccoli cv. "Legacy," DPPH and ABTS radical scavenging activity increased in samples treated to 20 μmol m^{-2} s^{-1} continuous white and blue LED light than control samples (Hasperué, Guardianelli et al., 2016). In addition, DPPH and ABTS radical scavenging activity were higher in outer Brussel sprouts after 5 days of treatment. In the inner leaves, there was no difference in antioxidant capacity until day 10 (Hasperué, Rodonia et al., 2016). Tomato fruit treated with additional UV and red light showed the highest hydrophilic and lipophilic antioxidant activity compared with fruits kept in darkness or treated with UV light only (Panjai et al., 2017).

Carotenoids

The highest content of carotenoids was detected in broccoli cv. "Legacy" heads under white and blue LED light during the storage (Hasperué, Guardianelli et al., 2016). The lycopene concentration increased in tomato cv. "Cappricia" additionally treated with UV and red light. Fruit treated with UV LED irradiation (60% UVB [280—320 nm with a dominant peak at 290 nm], 30% UVA [320—400 nm], 4% UVC [200—280 nm], and 6% visible light [400—700 nm] with red light [665 nm]) had the highest β-carotene concentration 15 days after harvest, in comparison with those radiated with only UV light or darkness. However, after 20 days of harvest, the β-carotene concentration was highest in control fruit (Panjai et al., 2017). Liu et al. (2009) reported that UVC treatment using germicidal lamps (254 nm, 13.7 kJ m^{-2}) increased the lycopene contents of breaker-stage tomato cv. "Red Ruby" by 1.4- and 1.8-fold at days 15 and 21 of storage compared with untreated fruit. The β-carotene content in UVC-treated tomato fruit did not change significantly during 21 days of treatment and storage.

To summarize, the effect of UV light on color and texture attributes and nutritional value in fruits and vegetables during postharvest depends on dose of irradiation, temperature, and storage duration. There are not many studies published on UVA LED lighting during postharvest storage; however, the effects can be similar to those of blue LED light due to common photoreceptory mechanisms.

Fungal Control

Fungal spoilage, quality damages, losses of nutritional value, and seed viability cause losses of approximately of the one-third of all produced food (about 1.3 billion tons) worldwide (Kumar & Kalita, 2017). During postharvest storage, various fungal pathogens are responsible for horticultural crop diseases and mycotoxins contamination. The main fungal plant pathogens during storage are various rots, molds, and mildews, which infect a wide range of plant species (Pétriacq, López, & Luna, 2018; Valiuškaitė, Kviklienė, Kviklys, & Lanauskas, 2006). The primary challenge for agricultural scientists remains to extend the shelf life of fruits and vegetables, due to their spoilage caused by the pathogens. Therefore, various physical and chemical strategies are applied to prolong and maintain the shelf life of horticultural crops. LEDs over traditional UV lamps use have an option to select specific wavelengths for targeted plant pathogen control (D'Souza et al., 2015; Hasan, Bashir, Ghosh, Lee, & Bae, 2017; Neugart & Schreiner, 2018).

The use of UV light offers direct and indirect plant protection that will reduce the use of chemical pesticides. However, the UV LED light technology in horticulture is a quite new research direction, and postharvest UV LED irradiation data are still lacking (D'Souza et al., 2015; Gordon, 2018; Hasan et al., 2017; Huché-Thélier et al., 2016; Latorre, Rojas, Díaz, & Chuaqui, 2012; Neugart & Schreiner, 2018; Wargent & Jordan, 2013).

There are only a few researches investigated the direct effect of UV LED light on plant pathogens; however, it depends on the wavelength of UV (Aihara et al., 2014; Britz et al., 2013; Lattore et al., 2012; Schuerger & Brown, 1997; Shirai, Watanabe, & Matsuki, 2016; Neugart & Schreiner, 2018) (Table 4.1). The research on UV light emitted by fluorescent, low-pressure vapor, and other lamps are much more available compared with studies using UV LED as a light source (Neugart & Schreiner, 2018; Schuerger & Brown, 1997; Wargent, Taylor, & Paul, 2006). The use of UVA and UVB applied by narrow-banded LED could decrease the use of repellents and natural opponents. It has been reported that UVA has effects on bacteria and pathogenic microbes (D'Souza et al., 2015; Hasan et al., 2017; Neugart & Schreiner, 2018). However, there are only several reports on UVA LED effect on pathogens, although major of them on foodborne bacteria (Aihara et al., 2014; Neugart & Schreiner, 2018; Shirai et al., 2016). In addition, the efficacy of UVA combined with proactive nanoparticles was greater than UVA alone (D'Souza et al., 2015). The photosterilizer chlorophyllin derivative and light (400 nm) reduced postharvest *Botrytis cinerea* on strawberry and raspberry (Rasiukevičiūtė et al., 2015).

Elad (1997) reported that UVA light stimulates the sporulation of some vegetable pathogens. The effects of blue LED along with UVA can vary depending on plants and pathogens. It was found that *Sphaerotheca fuliginea* on cucumber leaves was lowest under exposure to 660 nm red LED, but a number of colonies per leaf were highest under UVA + blue LED (Schuerger & Brown, 1997).

According to various researches, UVB radiation has high potential in plant protection against pathogens and pests (Ballaré, 2014; Neugart & Schreiner, 2018;

Table 4.1 The Effects of Ultraviolet (UV) and Blue (B) Light Conditions on Horticultural Crops During Postharvest and Fungus Control.

Plant and Light Treatments	Storage Conditions	Effect	Reference
Strawberry cv. "Sulyang"; UVA LED (385 nm). Control—nonirradiation	5°C, 4 D	Increased TSS content; No significant differences on TA, vitamin C, Anth, TPC.	Kim et al., 2011
Cucumber cv. "Silor"; UVA (GL) for 5, 10, 20 or 30 min. Control—dark.	5°C, 10°C, 15 D	Lower L^*, a^*, b^* values; Higher Hue angle (30 min), TSS (10°C); No difference on WL.	Kasim & Kasim, 2008
Tomato cv. "Red Ruby"; UVC (254 nm, 13.7 $kJ\,m^{-2}$, 5 min)	0, 4, 15, 21 D	Lower a^*, Hunter a^*/b^* values; Increased the Lyc content (15, 21 D); No difference on b^* value, β-Car.	Liu et al., 2009
Tomato cv. "Zinac"; UVC (254 nm, 0.32, 0.97, 2.56, 4.16 and 4.83 $kJ\,m^{-2}$ for 1, 3, 8, 13 and 15 min, respectively)	10 ± 0.5°C, 0, 9, 15 D	Lower a^* value; Higher h value; TPC (0.32 $kJ\,m^{-2}$, 9 D).	Pinheiro et al., 2015
Bell pepper cv. "Jubilee F1"; UVC (254 nm, 1.3, 2.4, and 3.6 $kJ\,m^{-2}$)	10 ± 0.5°C, 38 D	Higher WL (2.4 $kJ\,m^{-2}$), TA (1.3, 2.4 $kJ\,m^{-2}$); Lower WL (3.6 $kJ\,m^{-2}$), TSS, TA (3.6 $kJ\,m^{-2}$).	Karaşahin Yıldırım & Pekmezci, 2017
Tomato cv. "Cappricia"	20/19°C (day/night), 20 D	Decreased Flav and TPC (5 D); Higher AOX, Lyc content (UV + red), β-Car (UV, 15 D).	Panjai et al., 2017
Strawberry cv. "Fengguang"; B 470 nm (40 $\mu mol\,m^{-2}\,s^{-1}$)	5°C, 12 D	Increased CIRG value, TA, SOD, CAT catalase, APX activities, AA (4 D), TPC (2 D), DPPH radical scavenging activity (4 D); Lower levels of O_2^- production and H_2O_2 content (4 D), MDA level.	Xu et al., 2014
Cabbage cv. "Dongdori"; UVB LED	4–5°C, 18 D	Decreased L^*, a^*; Increased b^* values, VIT-C, TPC.	Lee et al., 2014

Continued

Table 4.1 The Effects of Ultraviolet (UV) and Blue (B) Light Conditions on Horticultural Crops During Postharvest and Fungus Control.—cont'd

Plant and Light Treatments	Storage Conditions	Effect	Reference
Broccoli cv. "Legacy"; B and W LED (20 µmol m^{-2} s^{-1})	5°C for 0, 35, and 42 D; 22°C for 0, 2, 3, and 4 D	Decreased L*, chlorophyll a and b content; Higher Hue angle, WL, Glu (2 D, 22°C), Fru (3 D, 22°C), Suc (4 D, 5 and 22°C), AA, DPPH and ABTS radical scavenging activity, Car.	Hasperué, Guardianelli et al., 2016
Brussel sprout; B and W LED (20 µmol m^{-2} s^{-1})	22°C, 10 D	Decreased L*; Higher WL, TSS (until 5 D, 22°C), Flav, DPPH and ABTS radical scavenging activity (5 D).	Hasperué, Rodonia et al., 2016
Tomato cv. "Dotaerang"; B 440–450 nm LED (85.7 µmol m^{-2} s^{-1}). Control—dark.	0, 3, 7 D	Increased a*, b* values; Decreased firmness.	Dhakal & Baek, 2014
Blueberry; B 480 nm LED (27 µW cm^{-2})	Not presented	No effect on L* value; Decreased TPC.	Spotts et al., 2017

		Pathogen	Effect	
Blueberry; UVA (361 nm), UVB (302 nm) and UVC (254 nm) at 40–110 mJ cm^{-2}	Petri plates kept at 20°C for 7 D.	B. cinerea, C. cladosporioides, C. herbarum, P. clavispora	Conidia inactivation	Lattore et al., 2012
Strawberry; deep UVC (272, 289, 293 nm, 20 mW m^{-2})	At 5°C	B. cinerea	Reduced mold growth	Britz et al., 2013
Tomatoes; B LED (150 µmol m^{-2} s^{-1})		B. cinerea	Inhibition of development	Kim et al., 2013
Lettuce; B LED (200 µmol m^{-2} s^{-1})		B. cinerea	Continuous irradiation suppressed development.	Kook et al., 2013
B LED (40 µmol m^{-2} s^{-1})		P. digitatum	Reduced the expansion	Liao et al., 2013; Alferez et al., 2012

AA, ascorbic acid; Anth, anthocyanin; AOX, antioxidant activity; APX, ascorbate peroxidase; B, blue; Car, carotenoids; CAT, catalase; CIRG, color index red grapes; D, day; Flav, flavonoids; Fru, fructose; GL, germicidal lamp; Glu, glucose; LED, light-emitting diode; Lyc, lycopene; MDA, maleic dialdehyde; SOD, superoxide dismutase; Suc, sucrose; TA, total acidity; TPC, total phenolic content; TSS, total soluble solids; UVA, ultraviolet A; UVB, ultraviolet B; UVC, ultraviolet C; VIT-C, vitamin C; W, white; WL, weight loss; β-Car, β-carotene.

Wargent, Jordan, 2013). UVB light subsequently induces the resistance to disease infection if the exposure time is extended (Wargent et al., 2006). Lettuce responses to UVB radiation showed a negative correlation to the contamination of *Bremia lactucae* and *B. cinerea* (Paul et al., 2012). The UVB radiation increases *Arabidopsis* resistance to *B. cinerea* (Demkura & Ballaré, 2012). For instance, Suthaparan et al. (2014) reported that UVB affects suppression of *Podosphaera xanthii* of cucumber. after daily application of UVB at intensity of 1 W m^{-2} the severity of *P. xanthii* significantly reduced to 15%. It is reported that night-time UV illumination affects *P. xanthii*. The exposure of UVB light during the day (to 16 h) and night time significantly reduced contamination of *P. xanthii* in cucumber (Suthaparan, Solhaug, Stensvand, & Gislerød, 2017).

Several researchers reported that UV light influences the development of pathogen conidia on plants. Latorre et al. (2012) found out that UV LED dose had significant effects on conidia survival. The results showed the reaction between fungal species and UV dose (UVB 302 nm and UVC 254 nm), but no significant play with interactions under UVA (361 nm). For instance, the adverse effect of UV LED was determined on the *B. cinerea* conidia survival in Petri plates. It was noticed that of all UV treatments, the highest inhibition of *B. cinerea* was approximately of 19% after UVC exposure. The *Cladosporium cladosporioides* and *C. herbarum* conidia were mostly resistant to UVA, UVB, and UVC LED at doses 40 and 110 mJ cm^{-2}. Lattore et al. (2012) observed that different UVA, UVB, and UVC irradiance inhibit growth of conidia of *B. cinerea*, *C. cladosporioides*, *C. herbarum*, and *Pestalotiopsis clavispora* on growth media. The interaction between fungal species and UV light dose were significant, but no significant interaction under UVA (361 nm). The most resistant to UV between 40 and 110 mJ cm^{-2} were *C. cladosporioides* and *C. herbarum* conidia (Lattore et al., 2012). However, red LED stimulates powdery mildew resistance (Wang et al., 2010).

Britz et al. (2013) explored the potential of deep UVC and UVB LED applications for a shelf-life extension during cold storage of fresh strawberry. The produce was exposed to low doses of UV LED (272, 289, or 293 nm) range (Table 4.1). The results showed that UV LED (272, 289, or 293 nm) at 20 mW m^{-2} reduced strawberry mold growth (suspected *B. cinerea*) up to 9 days compared with product stored in the dark. Paul et al. (2012) evaluated the UV radiation effect on lettuce pathogens *B. lactucae* and *B. cinerea* and found out that the disease severity of pathogens was reduced. The UVC treatment at 8 kJ m^{-2} significantly inhibited the growth of *Penicillium digitatum* on tomato fruit during 16-days storage at 16°C (Obande, Tucker, & Shama, 2011).

In several studies, reported the effect of UVC on tomato, Charles, Makhlouf, and Arul (2008a) performed research on postharvested tomato fruits with the dose of 3.7 kJ m^{-2} UVC to evaluate induced resistance to *B. cinerea*. It was found out that UVC-treated tomato fruits had lower *B. cinerea* surface colonization than that of control fruits. The UVC (dose 3.7 kJ m^{-2}) treatment of tomato fruits induced accumulation of phenolic compounds, such as lignin and suberin. These compounds develop the unsuitable environment for the fungal growth

(Charles, Goulet, & Arul, 2008b). The other obtained results showed a reduction of decay in UVC light−treated tomatoes. The infection of *Rhizopus stolonifer* in tomato after 72 h of UVC treatment was only 47% compared with control (Stevens et al., 2004). In addition, the UVC-treated tomato resistance to *B. cinerea* could be altered and proteins related to pathogenesis, as b-1,3-glucanase with glucanohydrolase activities are involved in disease resistance mechanism (Charles, Tano, Asselin, & Arul, 2009). In previous studies Liu et al. (1993) found that UVC doses from 1.3 up to 40 kJ m^{-2} can reduce contamination of *Alternaria alternata*, *B. cinerea*, and *R. stolonifer*, besides extending tomato' shelf life.

Several other researchers found out that in postharvest lettuce treated with germicidal UVC (254 nm) at dose 0.85 kJ m^{-2} resistance against *Sclerotinia minor* increased (Ouhibi et al., 2015) and *B. cinerea* (Ouhibi et al., 2015; Vasquez et al., 2017). Similarly, Mercier, Baka, Reddy, Corcuff, and Arul (2001) observed induced resistance to *B. cinerea* on bell pepper; however, the exposure to UVC at 0.88 kJ m^{-2} mostly reduced the number of natural infections occurring during storage.

Several combinations of UVC were applied to control *B. cinerea* on fruits. Nigro, Ippolito, Lattanzio, Venere, and Salerno (2000) investigated the germicidal UVC light of 0.25−4.0 kJ m^{-2} doses on postharvest strawberry contaminated with *B. cinerea*. Interestingly, the preharvest strawberry fruit treatment with UVC of 0.50 kJ m^{-2} and 1.00 kJ m^{-2} reduced concentrations of postharvest *B. cinerea* after artificial inoculations and natural infections, comparing with control. The results showed that significantly lower botrytis rot severity was found in fruits irradiated at 0.25 and 0.50 kJ m^{-2}. Pombo, Rosli, Martínez, and Civello (2011) also studied the induced resistance to *B. cinerea* in strawberry after both sides of fruits were irradiated with a dose of 4.1 kJ m^{-2} UVC. Preharvest UVC treatment of strawberries showed that after 9 days of storage only 30% of fruits had signs of infection. It was reported that the UVC irradiation increased the expression and activity of several enzymes involved in pathogen defense mechanism.

Similarly, Erkan et al. (2008) research showed that UVC at 0.43, 2.15, and 4.30 kJ m^{-2} doses reduced decay severity, besides increased antioxidant capacity and enzyme activities in strawberry fruit. The UVC (280 nm) treatments from 5 up to 60 min can reduce pathogen population on seeds of mung bean cv. "Wilczek." In addition, the root infections caused by *Fusarium* spp., *R. solani*, and *M. phaseolina* were reduced after 5, 10, 15, and 20 min (Siddiqui, Dawar, Zaki, & Hamid, 2011).

It was also reported that blue LED light could suppress the germination and sporulation of *Botrytis* spp., *Penicillium* spp., *Aspergillus* spp., *Phomopsis* spp. and far-red, red, and blue could inhibit *Aspergillus* spp. and other pathogens (Ballaré, 2014; Kim et al., 2013; Kook et al., 2013; Liao, Alferez, & Burns, 2013). Although the effect of blue LED was confirmed, the question was raised about the intensity and duration of the light. Kim et al. (2013) reported that 50−150 µmol m^{-2} s^{-1} blue LED induced *B. cinerea* resistance in tomato. In addition, Kook et al. (2013) observed that 200 µmol m^{-2} s^{-1} blue LED induced resistance in lettuce.

Liao et al. (2013) reported that blue LED exposure for 12 and 24 h significantly reduced the expansion of *P. digitatum* infection. Similarly, Alferez, Liao, and Burns (2012) reported that 12 h (followed by 12 h of dark period) treatment of blue LED (410 and 540 nm) per day were effective reducing *P. digitatum* mycelium. Therefore, when choosing the exposure time for any specific product it would be worthy to consider economically efficient time. Based on available data, in an overview of Hasan et al. (2017) it was concluded that red, blue, and green LED light could induce systemic resistance to fungal pathogens.

The effect of different wavelengths (white, red, green, blue) of fluorescent lamps on disease development varied, as for example the mold on tomatoes mostly developed in the dark. However, it was noticed that mold development on tomato fruit surface was not synchronized with rot infection with different light conditions in fruits treated with red and white light. *B. cinerea* pathosystems developed less fluffy mold than in the dark condition despite that red, white, and dark conditions caused similar soft decay areas (Zhu et al., 2013). Wang et al. (2010) reported that red LED enhanced resistance to *S. fuliginea* in cucumber. Kasim and Kasim (2017) evaluated the decay of lettuce during storage under different LED light (white, green, red, blue) at the intensity of 10 μmol m^{-2} s^{-1}. They found that the red LED light was most effective during 21-day storage, but the shelf life improved better under green and white LED. The content of lettuce leaf chlorophyll (SPAD) increased by white LED light treatment during both storage and shelf life conditions. The new research has started on the influence of the light quantity and quality to the strawberry *Botrytis* spp. and *Colletotrichum* spp. bioecologic properties in vitro and in vivo. The primary results shows that royal blue 455 nm, blue 470 nm, cyan 505 nm, yellow 590 nm, and red 627 nm LED light affect *B. cinerea* morphologic and phenotypic characteristic. It was noticed that cyan LED (505 nm) influenced *B. cinerea* growth in 16, 20, and 24 h photoperiods. However, the research is in the initial stage and needs further investigations (Rasiukevičiūtė et al., 2018).

Conclusions

Overview of reported scientific literature showed that application of visible and UV irradiation at pre- and postharvest stage stimulates synthesis and accumulation of health-beneficial bioactive compounds, extend shelf life of fresh horticulture products, and prevent against fungal spoilage. However, UV response is related to interactions between various environmental factors such as dose, exposure time, or storage temperature and other aspects such as cultivar, production type, pathogen, etc. For the reason of relatively low technology cost, most studies have been done with UVC irradiation on horticultural plants, whereas UVA and UVB irradiation is mostly effective when used for few hours or days at preharvest stage or during storage, and more specific equipment for this purpose are required. Literature data also showed that fluorescent tubes, mercury, and pulsed xenon lamps are dominant artificial light sources for UV treatments related to postharvest preservation and produce storage. However, these fixtures are space-consuming and have a wide range of wavelengths. Developing UV LED technology creates opportunity to select specific wavebands and their combination for the purpose to enhance and prolong postharvest quality of horticultural products. On the other hand, in case of LED applications, it is easier to interact with various environmental factors, storage technologies, and to install in different storage areas. However, because of quite short shelf life and relatively high cost comparing with other LEDs, uses of UV LED in horticultural technologies are under development, and limited studies are presented in this area.

Acknowledgments

This research was funded by the European Social Fund under the No 09.3.3-LMT-K-712 "Development of Competences of Scientists, other Researchers and Students through Practical Research Activities" measure project No 09.3.3-LMT-K-712-02-0052.

References

Aiamla-or, S., Kaewsuksaeng, S., Shigyo, M., & Yamauchi, N. (2010). Impact of UV-B irradiation on chlorophyll degradation and chlorophyll-degrading enzyme activities in stored broccoli (*Brassica oleracea* L. Italica Group) florets. *Food Chemistry, 120*, 645–651.

Aihara, M., Lian, X., Shimohata, T., Uebanso, T., Mawatari, K., Harada, Y., et al. (2014). Vegetable surface sterilization system using UVA light-emitting diodes. *Journal of Medical Investigation, 61*(3–4), 285–290.

Alferez, F., Liao, H.-L., & Burns, J. K. (2012). Blue light alters infection by *Penicillium digitatum* in tangerines. *Postharvest Biology and Technology, 63*, 11–15.

Arias, R., Lee, T. C., Logendra, L., & Janes, H. (2000). Correlation of lycopene measured by HPLC with the L*, a*, b* colour readings of a hydroponic tomato and the relationship of maturity with colour and lycopene content. *Journal of Agricultural and Food Chemistry, 48*, 1697–1702.

Artés-Hernández, F., Escalona, V. H., Robles, P. A., Martínez-Hernández, G. B., & Artés, F. (2009). Effect of UV-C radiation on quality of minimally processed spinach leaves. *Journal of the Science of Food and Agriculture, 89*, 414–421.

Ballaré, C. L. (2014). Light regulation of plant defense. *Annual Review of Plant Biology, 65*, 335–363.

Bantis, F., Smirnakou, S., Ouzounis, T., Koukounaras, A., Ntagkas, N., & Radoglou, K. (2018). Current status and recent achievements in the field of horticulture with the use of light-emitting diodes (LEDs). *Scientia Horticulturae, 235*, 437–451.

Barka, E. A. (2001). Protective enzymes against reactive oxygen species during ripening of tomato (*Lycopersicon esculentum*) fruits in response to low amounts of UV-C. *Australian Journal of Plant Physiology, 28*, 785–791.

Bian, Z. H., Yang, Q. C., & Liu, W. K. (2015). Effects of light quality on the accumulation of phytochemicals in vegetables produced in controlled environments: A review. *Journal of the Science of Food and Agriculture, 95*, 869–877.

Bjorkman, M., Klingen, I., Birch, A. N. E., Bones, A. M., Bruce, T. J. A., Johansen, T. J., et al. (2011). Phytochemicals of Brassicaceae in plant protection and human health—influences of climate, environment and agronomic practice. *Phytochemistry, 72*, 538—556.

Brazaityte, A., Sakalauskiene, S., Virsile, A., Jankauskiene, J., Samuoliene, G., Sirtautas, R., et al. (2016). The effect of short-term red lighting on Brassicaceae microgreens grown indoors. *Acta Horticulturae, 1123*, 177—184.

Brazaitytė, A., Viršilė, A., Jankauskienė, J., Sakalauskienė, S., Samuolienė, G., Sirtautas, R., et al. (2015). Effect of supplemental UV-A irradiation in solid-state lighting on the growth and phytochemical content of microgreens. *International Agrophysics, 29*(1), 13—22.

Britz, S., Gaska, I., Shturm, I., Bilenko, Y., Shatalov, M., & Gaska, R. (2013). Deep ultraviolet (DUV) light-emitting diodes (LEDs) to maintain freshness and phytochemical composition during postharvest storage. In *CLEO: 2013, OSA technical digest (online) (Optical Society of America, 2013), paper ATh3N.3*.

Carvalho, S. D., & Folta, K. M. (2014). Environmentally modified organisms—Expanding genetic potential with light. *CRC Critical Reviews In Plant Sciences, 33*, 486—508.

Chairat, B., Nutthachai, P., & Varit, S. (2013). Effect of UV-C treatment on chlorophyll degradation, antioxidant enzyme activities and senescence in Chinese kale (*Brassica oleracea* var. alboglabra). *International Food Research Journal, 20*(2), 623—628.

Charles, M. T., & Arul, J. (2007). UV treatment of fresh fruits and vegetables for improved quality: A status report. *Stewart Postharvest Review, 3*(6), 1—8.

Charles, M. T., Goulet, A., & Arul, J. (2008b). Physiological basis of UV-C induced resistance to *Botrytis cinerea* in tomato fruit IV. Biochemical modification of structural barriers. *Postharvest Biology and Technology, 47*, 41—53.

Charles, M. T., Makhlouf, J., & Arul, J. (2008a). Physiological basis of UV-C induced resistance to *Botrytis cinerea* in tomato fruit III. Ultrastructural modifications and their impact on fungal colonization. *Postharvest Biology and Technology, 47*, 27—40.

Charles, M. T., Tano, K., Asselin, A., & Arul, J. (2009). Physiological basis of UV-C induced resistance to *Botrytis cinerea* in tomato fruit. V. Constitutive defence enzymes and inducible pathogenesis related proteins. *Postharvest Biology and Technology, 51*, 414—424.

Costa, L., Vicente, A. R., Civello, P. M., Chaves, A. R., & Martínez, G. A. (2006). UV-C treatment delays postharvest senescence in broccoli florets. *Postharvest Biology and Technology, 39*, 204—210.

D'Souza, C., Yuk, H., Khoo, H. G., & Zhou, W. (2015). Application of light-emitting diodes in food production, postharvest preservation, and microbiological food safety. *Comprehensive Reviews in Food Science and Food Safety, 14*, 719—740.

De Vries, F. P. (1975). The cost of maintenance processes in plant cells. *Annals of Botany, 39*(1), 77—92.

Demkura, P. V., & Ballaré, C. L. (2012). UVR8 mediates UV-B-induced Arabidopsis defense responses against *Botrytis cinerea* by controlling sinapate accumulation. *Molecular Plant, 5*, 642—652.

Demotes-Mainard, S., Péron, T., Corot, A., Bertheloot, J., Le Gourrierec, J., Pelleschi-Travier, S., et al. (2016). Plant responses to red and far-red lights, applications in horticulture. *Environmental and Experimental Botany, 121*, 4—21.

Dhakal, R., & Baek, K.-H. (2014). Metabolic alternation in the accumulation of free amino acids and γ-aminobutyric acid in postharvest mature green tomatoes following irradiation with blue light. *Horticulture, Environment, and Biotechnology, 55*(1), 36—41.

D'Souza, C., Yuk, H. G., Khoo, G. H., & Zhou, W. (2015). Application of light-emitting diodes in food production, postharvest preservation, and microbiological food safety. *Comprehensive Reviews in Food Science and Food Safety, 14*, 719—740.

Elad, Y. (1997). Effect of filtration of solar light on the production of conidia by field isolates of *Botrytis cinerea* and on several diseases of greenhouse-grown vegetables. *Crop Protection, 16*, 635—642.

Erkan, M., Wang, S. Y., & Wang, C. Y. (2008). Effect of UV treatment on antioxidant capacity, antioxidant enzyme activity and decay in strawberry fruit. *Postharvest Biology and Technology, 48*, 163—171.

Fonseca, J. M., & Rushing, J. W. (2008). Application of ultraviolet light during postharvest handling of produce: Limitations and possibilities. *Fresh Produce, 2*(2), 41—46.

Funamoto, Y., Yamauchi, N., Shigenaga, T., & Shigyo, M. (2002). Effects of heat treatment on chlorophyll degrading enzymes in stored broccoli (*Brassica oleracea* L.). *Postharvest Biology and Technology, 24*, 163—170.

Gordon, P. (2018). UV-C LEDs for food safety. *IUVA News, 20*(2), 4—7.

Goto, E., Hayashi, K., Furuyama, S., Hikosaka, S., & Ishigami, Y. (2016). Effect of UV light on phytochemical accumulation and expression of anthocyanin biosynthesis genes in red leaf lettuce. *Acta Horticulturae, 1134*, 179—185.

Harbaum-Piayda, B., Walter, B., Bengtsson, G. B., Hubbermann, E. M., Bilger, W., & Karin Schwarz, K. (2010). Influence of pre-harvest UV-B irradiation and normal or controlled atmosphere storage on flavonoid and hydroxycinnamic acid contents of pak choi (*Brassica campestris* L. ssp. chinensis var. communis). *Postharvest Biology and Technology, 56*, 202—208.

Hasan, M. M., Bashir, T., Ghosh, R., Lee, S. K., & Bae, H. (2017). An Overview of LEDs' effects on the production of bioactive compounds and crop quality. *Molecules, 22*(9), 1420.

Hasperué, J. H., Guardianelli, L., Rodoni, L. M., Chaves, A. R., & Martínez, G. A. (2016). Continuous white-blue LED light exposition delays postharvest senescence of broccoli. *Lebensmittel-Wissenschaft und- Technologie- Food Science and Technology, 65*, 495—502.

Hasperué, J. H., Rodonia, L. M., Guardianelli, L. M., Chaves, A. R., & Martínez, G. A. (2016). Use of LED light for Brussels sprouts postharvest conservation. *Scientific Horticulture, 213*, 281—286.

Hewett, E. W. (2006). An overview of preharvest factors influencing postharvest quality of horticultural products. *International Journal of Postharvest Technology and Innovation, 1*(1), 4—15.

Hörtensteiner, S., & Kräutler, B. (2011). Chlorophyll breakdown in higher plants. *Biochimica et Biophysica Acta, 1807*, 977—988.

Huché-Thélier, L., Crespel, L., Gourrierec, J. L., Morel, P., Sakr, S., & Leduc, N. (2016). Light signaling and plant responses to blue and UV radiations - perspectives for applications in horticulture. *Environmental and Experimental Botany, 121*, 22—38.

Ilić, Z. S., & Fallik, E. (2017). Light quality manipulation improves vegetable quality at harvest and postharvest: A review. *Environmental and Experimental Botany, 139*, 79—90.

Jagadeesh, S. L., Charles, M. T., Gariepy, Y., Goyette, B., Raghavan, G. S. V., & Vigneault, C. (2011). Influence of postharvest UV-C hormesis on the bioactive components of tomato during post-treatment handling. *Food and Bioprocess Technology, 4*, 1463—1472.

Jansen, M. A. K., Hectors, K., O'Brien, N. M., Guisez, Y., & Potters, G. (2008). Plant stress and human health: Do human consumers benefit from UV-B acclimated crops? *Plant Science, 175*, 449–458.

Karaşahin Yildirim, I. K., & Pekmezci, M. (2017). Postharvest ultraviolet-C (UV-C) treatment reduces decay and maintains quality of bell peppers. *Akademik Ziraat Dergisi, 6*(2), 89–94.

Kasim, M., & Kasim, R. (2008). UV-A treatment delays yellowing of cucumber during storage. *Journal of Food Agriculture and Environment, 6*, 29–32.

Kasim, R., & Kasim, M. U. (2012). UV-C treatments on fresh-cut garden cress (*Lepidium sativum* L.) enhanced chlorophyll content and prevent leaf yellowing. *World Applied Sciences Journal, 17*(4), 509–515.

Kasim, M. U., & Kasim, R. (2017). While continuous white LED lighting increases chlorophyll content (SPAD), green LED light reduces the infection rate of lettuce during storage and shelf-life conditions. *Journal of Food Processing and Preservation, 41*(6), e13266.

Kim, K., Kook, H.-S., Jang, Y.-J., Lee, W.-H., Kamala-Kannan, S., Chae, J.-C., et al. (2013). The effect of blue-light emitting diodes on antioxidant properties and resistance to *Botrytis cinerea* in tomato. *Journal of Plant Pathology & Microbiology, 4*(9), 203.

Kim, B. S., Lee, H. O., Kim, J. Y., Kwon, K. H., Cha, H. S., & Kim, J. H. (2011). An effect of light emitting diode (LED) irradiation treatment on the amplification of functional components of immature strawberry. *Horticulture, Environment, and Biotechnology, 52*(1), 35–39.

King, G. A., & Morris, S. C. (1994). Early compositional changes during postharvest senescence of broccoli. *Journal of the American Society for Horticultural Science, 119*, 1000–1005.

Kinoshita, T., Doi, M., Suetsugu, N., Kagawa, T., Wada, M., & Shimazaki, K. I. (2001). Phot1 and phot2 mediate blue light regulation of stomatal opening. *Nature, 414*(6864), 656–660.

Konopacka, D., & Plocharski, W. J. (2003). Effect of the storage conditions on the relationship between apple firmness and texture acceptability. *Postharvest Biology and Technology, 32*, 205–211.

Kook, H. S., Park, S. H., Jang, Y. J., Lee, G. W., Kim, J. S., Kim, H. M., et al. (2013). Blue LED (light-emitting diodes)-mediated growth promotion and control of Botrytis disease in lettuce. *Acta Agriculturae Scandinavica, Section B — Soil & Plant Science, 63*(3), 271–277.

Kumar, D., & Kalita, P. (2017). Reducing postharvest losses during storage of grain crops to strengthen food security in developing countries. *Foods, 6*(1), e8.

Latorre, B. A., Rojas, S., Díaz, G. A., & Chuaqui, H. (2012). Germicidal effect of UV light on epiphytic fungi isolated from blueberry. *Ciencia e Investigacian Agraria, 39*(3), 473–480.

Lee, Y. J., Ha, J. Y., Oh, J. E., & Cho, M. S. (2014). The effect of LED irradiation on the quality of cabbage stored at a low temperature. *Food Science and Biotechnology, 23*(4), 1087–1093.

Lee, N. Y., Lee, M. J., Kim, Y. K., Park, J. C., Park, H. K., Choi, J. S., et al. (2010). Effect of light emitting diode radiation on antioxidant activity of barley leaf. *Journal of the Korean Society for Applied Biological Chemistry, 53*, 658–690.

Liao, H. L., Alferez, F., & Burns, J. K. (2013). Assessment of blue light treatments on citrus postharvest diseases. *Postharvest Biology and Technology, 81*, 81–88.

Li, Q., & Kubota, C. (2009). Effects of supplemental light quality on growth and phytochemicals of baby leaf lettuce. *Environmental and Experimental Botany, 67*(1), 59–64.

Lin, C., & Todo, T. (2005). The cryptochromes. *Genome Biology, 6*(5), 220.

Liu, C., Cai, L., Lu, X., Han, X., & Ying, T. (2012). Effect of postharvest UV-C irradiation on phenolic compound content and antioxidant activity of tomato fruit during storage. *Journal of Integrative Agriculture, 11*(1), 159−165.

Liu, J., Stevens, C., Khan, V., Lu, J., Wilson, C., Adeyeye, O., et al. (1993). Application of ultraviolet-C light on storage rots and ripening of tomatoes. *Journal of Food Protection, 56*, 868−873.

Liu, L. H., Zabaras, D., Bennett, L. E., Aguas, P., & Woonton, B. W. (2009). Effects of UV-C, red light and sun light on the carotenoid content and physical qualities of tomatoes during post-harvest storage. *Food Chemistry, 115*, 495−500.

Matile, P., Hörtensteiner, S., & Thomas, H. (1999). Chlorophyll degradation. *Annual Review of Plant Physiology and Plant Molecular Biology, 50*, 67−95.

Ma, L., Zhang, M., Bhandari, B., & Gao, Z. (2017). Recent developments in novel shelf life extension technologies of fresh-cut fruits and vegetables. *Trends in Food Science & Technology, 64*, 23−38.

Ma, G., Zhang, L., Setiawan, K. C., Yamawaki, K., Asai, T., Nishikawa, F., et al. (2014). Effect of red and blue LED light irradiation on ascorbate content and expression of genes related to ascorbate metabolism in postharvest broccoli. *Postharvest Biology and Technology, 94*, 97−103.

Mercier, J., Baka, M., Reddy, B., Corcuff, R., & Arul, J. (2001). Shortwave ultraviolet irradiation for control of decay caused by *Botrytis cinerea* in bell pepper: Induced resistance and germicidal effects. *Journal of the American Society for Horticultural Science, 126*(1), 128−133.

Mitchell, C. A., Dzakovich, M. P., Gomez, C., Burr, J. F., Hernández, R., Kubota, C., et al. (2015). Light-emitting diodes in horticulture. In J. Janick (Ed.), *Horticultural reviews* (1st ed., Vol. 43). Wiley-Blackwell. Published 2015 by John Wiley & Sons, Inc.

Neugart, S., & Schreiner, M. (2018). UVB and UVA as eustressors in horticultural and agricultural crops. *Scientific Horticulture, 234*(18), 370−381.

Nigro, F., Ippolito, A., Lattanzio, V., Venere, D. D., & Salerno, M. (2000). Effect of ultraviolet-c light on postharvest decay of strawberry. *Journal of Plant Pathology, 82*(1), 29−37.

Noichinda, S., Bodhipadma, K., Mahamontri, C., Narongruk, T., & Ketsa, S. (2007). Light during storage prevents loss of ascorbic acid, and increases glucose and fructose levels in Chinese kale (*Brassica oleracea* var. alboglabra). *Postharvest Biology and Technology, 44*(3), 312−315.

Obande, M. A., Tucker, G. A., & Shama, G. (2011). Effect of preharvest UV-C treatment of tomatoes (Solanum lycopersicon Mill.) on ripening and pathogen resistance. *Postharvest Biology and Technology, 62*, 188−192.

Olle, M., & Viršile, A. (2013). The effects of light-emitting diode lighting on greenhouse plant growth and quality. *Agricultural and Food Science, 22*(2), 223−234.

Ouhibi, C., Attia, H., Nicot, P., Lecompte, F., Vidal, V., Lachaâl, M., et al. (2015). Effects of nitrogen supply and of UV-C irradiation on the susceptibility of *Lactuca sativa* L. to *Botrytis cinerea* and *Sclerotinia minor*. *Plant and Soil, 393*, 35−46.

Panjai, L., Noga, G., Fiebig, A., & Hunsche, M. (2017). Effects of continuous red light and short daily UV exposure during postharvest on carotenoid concentration and antioxidant capacity in stored tomatoes. *Scientific Horticulture, 226*, 97−103.

Paul, N. D., Moore, J. P., McPherson, M., Lambourne, C., Croft, P., Heaton, J. C., et al. (2012). Ecological responses to UV radiation: Interactions between the biological effects of UV on plants and on associated organisms. *Physiologia Plantarum, 145*(4), 565−581.

Pérez-Ambrocio, A., Guerrero-Beltrán, J. A., Aparicio-Fernández, X., Ávila-Sosa, R., Hernández-Carranza, P., Cid-Pérez, S., et al. (2018). Effect of blue and ultraviolet-C light irradiation on bioactive compounds and antioxidant capacity of habanero pepper (*Capsicum chinense*) during refrigeration storage. *Postharvest Biology and Technology, 135*, 19–26.

Pétriacq, P., López, A., & Luna, E. (2018). Fruit decay to diseases: Can induced resistance and priming help? *Plants, 7*(4), E77.

Pinheiro, J., Alegria, C., Abreu, M., Gonçalves, E. M., & Silva, C. L. M. (2015). Use of UV-C postharvest treatment for extending fresh whole tomato (*Solanum lycopersicum*, cv. Zinac) shelf-life. *Journal of Food Science & Technology, 52*(8), 5066–5074.

Pombo, M. A., Rosli, H. G., Martínez, G. A., & Civello, P. M. (2011). UV-C treatment affects the expression and activity of defense genes in strawberry fruit (*Fragaria × ananassa*, Duch.). *Postharvest Biology and Technology, 59*, 94–102.

Rasiukevičiūtė, N., Bylaitė, A., Brazaitytė, A., Valiuškaitė, A., Vaštakaitė, V., Viršilė, A., et al. (2018). The influence of LED light on *Botrytis cinerea* biometric and biological features. In *International conference of the scientific actualities and innovations in horticulture 2018* (pp. 16–17).

Rasiukevičiūtė, N., Valiuškaitė, A., Uselis, N., Buskienė, L., Viškelis, J., & Lukšienė, Ž. (2015). New non-chemical postharvest technologies reducing berry contamination. *Zemdirbyste-Agriculture, 104*(4), 411–416.

Rouphaela, Y., Kyriacou, M. C., Petropoulos, S. A., Pascale, S., & Coll, G. (2018). Improving vegetable quality in controlled environments. *Scientia Horticulturae, 234*, 275–289.

Rybarczyk-Plonska, A., Hansen, M. K., Wolda, A.-B., Hagenb, S. F., Borge, G. I. A., & Bengtsson, G. B. (2014). Vitamin C in broccoli (*Brassica oleracea* L. var. italica) flower buds as affected by postharvest light, UV-B irradiation and temperature. *Postharvest Biology and Technology, 98*, 82–89.

Samuolienė, G., Brazaitytė, A., Jankauskienė, J., Viršilė, A., Sirtautas, R., Novičkovas, A., et al. (2013). LED irradiance level affects growth and nutritional quality of Brassica microgreens. *Central European Journal of Biology, 8*(12), 1241–1249.

Schuerger, A. C., & Brown, C. S. (1997). Spectral quality affects disease development of three pathogens on hydroponically grown plants. *HortScience, 32*, 96–100.

Serrano, M., Martinez-Romero, D., Guillen, F., Castillo, S., & Valero, D. (2006). Maintenance of broccoli quality and functional properties during cold storage as affected by modified atmosphere packaging. *Postharvest Biology and Technology, 39*(1), 61–68.

Shirai, A., Watanabe, T., & Matsuki, H. (2016). Inactivation of foodborne pathogenic and spoilage micro-organisms using ultraviolet-A light in combination with ferulic acid. *Letters in Applied Microbiology, 64*, 96–102.

Siddiqui, A., Dawar, S., Zaki, M. J., & Hamid, N. (2011). Role of ultra violet (UV-C) radiation in the control of root infecting fungi on groundnut and mung bean. *Pakistan Journal of Botany, 43*, 2221–2224.

Spotts, I., Sandhu, R. K., Singh, A., & Collier, C. M. (2017). Design of light emitting diode system for postharvest shelf-life enhancement of fresh produce. In *CSBE/SCGAB 2017 annual conference Canada Inns Polo Park, Winnipeg, MB, 6–10 August 2017*.

Stevens, C., Liu, J., Khan, V., Lu, J., Kabwe, M., Wilson, C., et al. (2004). The effects of low-dose ultraviolet light-C treatment on polygalacturonase activity, delay ripening and Rhizopus soft rot development of tomatoes. *Crop Protection, 23*, 551–554.

Suthaparan, A., Solhaug, K. A., Stensvand, A., & Gislerød, H. R. (2017). Daily light integral and day light quality: Potentials and pitfalls of nighttime UV treatments on cucumber powdery mildew. *Journal of Photochemistry and Photobiology B, 175*, 141–148.

Suthaparan, A., Stensvand, A., Solhaug, K. A., Torre, S., Telfer, K. H., Ruud, A. K., et al. (2014). Suppression of cucumber powdery mildew by supplemental UV-B radiation in greenhouses can be augmented or reduced by background radiation quality. *Plant Disease, 98*, 1349—1357.

Tomás-Callejas, A., Otón, M., Artés, F., & Artés-Hernández, F. (2012). Combined effect of UV-C pretreatment and high oxygen packaging for keeping the quality of fresh-cut tatsoi baby leaves. *Innovative Food Science & Emerging Technologies, 14*, 115—121.

Topcu, Y., Dogan, A., Kasimoglu, Z., Sahin-Nadeem, H., Polat, E., & Erkan, M. (2015). The effects of UV radiation during the vegetative period on antioxidant compounds and post-harvest quality of broccoli (*Brassica oleracea* L.). *Plant Physiology and Biochemistry, 93*, 56—65.

Urban, L., Charles, F., de Miranda, M. R. A., & Aarrouf, J. (2016). Understanding the physiological effects of UV-C light and exploiting its agronomic potential before and after harvest. *Plant Physiology and Biochemistry, 105*, 1—11.

Valiuškaitė, A., Kviklienė, N., Kviklys, D., & Lanauskas, J. (2006). Post-harvest fruit rot incidence depending on apple maturity. *Agronomy Research, 4*(Special issue), 427—431.

Vaštakaitė, V., Viršilė, A., Brazaitytė, A., Samuolienė, G., Jankauskienė, J., Novičkovas, A., et al. (2017). Pulsed light-emitting diodes for higher phytochemical level in microgreens. *Journal of Agricultural and Food Chemistry, 65*(31), 6529—6534.

Vaštakaitė, V., Viršilė, A., Brazaitytė, A., Samuolienė, G., Jankauskienė, J., Sirtautas, R., et al. (2016). The effect of UV-A supplemental lighting on antioxidant properties of *Ocimum basilicum* L. microgreens in greenhouse. In *Proceedings of the 7th international scientific conference rural development 2015*.

Wang, H., Jiang, Y. P., Yu, H. J., Xia, X. J., Shi, K., Zhou, Y. H., et al. (2010). Light quality affects incidence of powdery mildew, expression of defence-related genes and associated metabolism in cucumber plants. *European Journal of Plant Pathology, 127*, 125—135.

Wargent, J. J. (2016). UV LEDs in horticulture: From biology to application. *Acta Horticulturae, 1134*, 25—32.

Wargent, J. J., & Jordan, B. R. (2013). From ozone depletion to agriculture: Understanding the role of UV radiation in sustainable crop production. *New Phytologist, 197*, 1058—1076.

Wargent, J. J., Taylor, A., & Paul, N. D. (2006). UV supplementation for growth regulation and disease control. *Acta Horticulturae, 711*, 333—338.

Wiczkowski, W., Szawara-Nowak, D., & Topolska, J. (2013). Red cabbage anthocyanins: Profile, isolation, identification, and antioxidant activity. *Food Research International, 51*, 303—309.

Witkowska, I. M. (2013). *Factors affecting the postharvest performance of fresh-cut lettuce* (Ph.D. thesis). Wageningen, NL: Wageningen University.

Wu, J., Liu, W., Yuan, L., Guan, W.-Q., Brennan, C. S., Zhang, Y.-Y., et al. (2017). The influence of postharvest UV-C treatment on anthocyanin biosynthesis in fresh-cut red cabbage. *Scientific Reports, 7*, 5232.

Xu, F., Shi, L., Chen, W., Cao, S., Su, X., & Yang, B. (2014). Effect of blue light treatment on fruit quality, antioxidant enzymes and radical-scavenging activity in strawberry fruit. *Scientific Horticulture, 175*, 181—186.

Zhu, P., Zhang, C., Xiao, H., Wang, Y., Toyoda, H., & Xu, L. (2013). Exploitable regulatory effects of light on growth and development of *Botrytis cinerea. Journal of Plant Pathology, 95*(3), 509—517.

UV Light-Emitting Diodes (LEDs) and Food Safety

Tatiana Koutchma, PhD [1]**, Vladimir Popović, MSc** [2]

Research Scientist, Guelph Research and Development Center, Agriculture and Agri-Food Canada, Guelph, ON, Canada[1]*; Guelph Research and Development Center, Agriculture and Agri-Food Canada, Guelph, ON, Canada*[2]

Chapter outline

UV Light for Water Disinfection and Food Safety 92
Concept of Photoinactivation Mechanism of UV light.. 93
    UVA 94
    UVB 95
    UVC 96
    Blue Light.. 97
Action Spectra for Water- and Foodborne Pathogens 99
    Waterborne Pathogens 99
    Foodborne Pathogens.. 100
    Multiple Wavelengths Application of UV and Visible LEDs............................. 105
    LEDs for Food Safety Applications 105
    UV LEDs for Fresh Produce 109
    Blue and Visible Light LEDs 111
    Effect of UV LEDs on Food Quality and Nutritional Parameters...................... 111
Conclusions... 113
References 114
Further Reading 117

Abstract

Safety of food processing facilities and finished products can greatly benefit from application of UV light technologies to control microbial hazards through treatments of air, nonfood and food contact surfaces, ingredients, packaging, raw and finished products. Low-pressure lamps emitting light at 253.7 nm are currently approved as a main radiation sources for treatment of juices, surfaces, or vitamin D generation with the limitations in terms of ozone production. Currently, UV LED technology is quickly extending in the UVB and UVC germicidal range as alternative to traditional UV lamps. Despite the low efficiency of UVB and UVC LEDs (approximately at 4%), LEDs are on growing trend and UVA LEDs are already being incorporated into point-of-use units to serve the defense and outdoor industries and tested for water treatment. Food safety application is the next step of UV LED adaption by industry. This chapter will review the germicidal action of UVA, UVB, UVC, and blue light, existing research, and first reported application of UV LEDs against foodborne pathogens at multiple wavelengths. Understanding of bacterial action spectra is also discussed to provide basis for optimization of the most effective treatment using single or wavelengths combinations. Considerations for UV LED treatment of fresh produce with some reported effects on quality and nutritional attributes are included.

Keywords: Food applications; Food pathogens action spectra; Multiple wavelengths; Safety; Ultraviolet (UV) light; UV light-emitting diodes (UVC, UVA LEDs); UVB.

UV Light for Water Disinfection and Food Safety

Ultraviolet (UVC) disinfection of water was introduced in Europe in 1930 and continued to dominate in Europe for drinking water treatment until the end of 20 century. In the United States, UV was used mainly for wastewater and changed when it was discovered that UVC light was not only effective against bacteria and viruses, but it also could prevent the spread of *Cryptosporidium* and *Giardia*, protozoan pathogens that are highly resistant to chlorine. By 2006, the US Environmental Protection Agency (EPA) required a secondary barrier, in addition to chlorine, to prevent *Cryptosporidium* and *Giardia* outbreaks (USEPA, 2006). The largest UV disinfection system in the world treats the water for New York City. A UV dose of 40 mJ cm^{-2} achieves 4-log (99.99%) inactivation of bacteria and most viruses. A high UV dose of 186 mJ cm^{-2} is required for a 4-log inactivation of adenovirus. For surface water supplies and ground water under the influence of surface water, a 40 mJ cm^{-2} UV device can be used in combination with chlorine to effectively inactivate viruses.

In the past two decades, food processors have been using UV treatments as practical solutions to produce microbiologically safe foods with better quality and more often with enhanced health benefits.

The US Food and Drug Administration (FDA) approved UV light as alternative treatment to thermal pasteurization of fresh juice products (US FDA, 2001). In 2001, the FDA amended the food additive regulations to provide for the safe use of UV radiation to reduce human pathogens and other microorganisms in juice products. Because UV dose necessary for human pathogen reduction depends on the type of juice, the initial microbial load, and the design of the irradiation system (e.g., flow rate, number of lamps, and time exposed to irradiation), the US FDA did not specify a minimum or maximum dose by regulation and the levels of UV dose applied to the juice will be limited by the possible alterations in organoleptic characteristics of the juice after UV irradiation.

Other approved applications of UV radiation for the processing and treatment of foods are surface microorganisms control and sterilization of potable water, shelf life extension of pasteurized milk, and formation of vitamin D in bread (EFSA, 2016; US FDA, 2012). Low-pressure lamps at 253.7 nm are approved as main radiation sources with the limitations in terms of ozone production.

Also, food processing facility can benefit from application of UV light technologies to control microbial hazards through treatments of air, nonfood, food contact surfaces, ingredients, and raw and finished products. Although there are a number of benefits for light-based technologies in food industry, cost-saving opportunities of energy, processing water, and enhanced safety need to be carefully considered in each specific case for successful technology implementation and to assure positive benefits.

Concept of Photoinactivation Mechanism of UV light

Bactericidal action of UV and blue light and near-infrared irradiation has been known for a long time. It has been reported that bacterial isolates have different susceptibilities to UV radiation and UV sensitivity is dependent on the wavelength and photon energy that are irreversibly proportional: the longer wavelength corresponds to the lower energy of photon and consequently the penetration depth.

Discussing the concept of photoinactivation by sun light (Nelson et al. 2018) identified the absorption of a photon by a chromophore or functional group that causes a conformational change of the molecule when hit by light, as a first step in any photoinactivation mechanism. Direct photoinactivation occurs when a chromophore endogenous to the microorganism (constituent of the microorganism, e.g., nucleic acids, proteins, or other macromolecules that occur in microorganisms) absorbs a photon, resulting in changes to the chemical structure of the chromophore. Indirect photoinactivation occurs when an endogenous or exogenous (not a constituent of the microorganism) chromophore absorbs a photon and sensitizes the production of photoproduced reactive intermediates (PPRI) that damage virus or cell components. Chromophores that produce PPRI are also called sensitizers. In viruses, chromophores involved in the endogenous direct and indirect inactivation are limited to amino acids (tryptophan, tyrosine, cysteine) and nucleic acid bases that primarily absorb light in the UVB range. In bacteria, chromophores also include coenzymes, vitamins, and metalloproteins. All three mechanisms may contribute to bacterial inactivation and likely occur simultaneously and interact in bacteria.

In case of direct inactivation, UV light photons are absorbed by the chemical bonds holding DNA complex helical structure in place. Because the different biomolecules (e.g., DNA, proteins, and lipids) absorb UV radiation at different wavelengths, there is a variability of cellular targets of various wavelengths that determines the wavelengths efficiency and UV dose. Jagger (1967) indicated that the inactivation dose required for comparable levels of inactivation at 400, 340, and 300 nm are nominally 10^4, 10^3, and 10 times higher, respectively, than that required at 260 nm.

DNA is considered the major target of UV radiation. However, comparable levels of DNA photoproduct accumulation are observed in bacteria displaying different sensitivities to UV radiation (Santos et al., 2013). In addition, DNA damage alone cannot account for the inhibition of bacterial activity because of the damage to other biomolecules that contribute to the inhibitory effects of UV radiation and consequently to the variation in UV sensitivity among bacterial isolates. Recently, Santos et al. (2013) reported the targets of UV radiation of different spectral regions UVA, UVB, and UVC across a range of bacteria with different UV susceptibilities. In particular, the involvement of DNA damage in eliciting bacterial inactivation upon UVA exposure was observed, which is in accordance with reports on the induction of the SOS response in UVA-irradiated bacteria. Oxidative damage to lipids was found to be determinant for bacterial inactivation during UVB exposure. Such observation is in agreement with recent reports of enhanced expression of the glyoxalase protein and alkyl hydroperoxide reductase AhpC, involved in the detoxification of lipid peroxidation by-products, following UVB exposure. Finally, oxidative stress was also found to be crucial for cell inactivation under UVC, supporting evidence accumulating in recent years. Summary of UV light germicidal action in UVA, UVB, and UVC range is presented in Table 5.1.

UVA

Limited research has been conducted on the effectiveness of UV LEDs for disinfection. These studies have shown that different water- and foodborne pathogens and spoilage microorganisms, as well as their individual strains, can vary in UV

Table 5.1 Summary of UV Light Germicidal Action in UVA, UVB, and UVC Range.

UV Diapason	Wavelength (nm)	Photon Energy (eV, aJ)	Germicidal Mechanism
UVA	320–400	3.10–3.94 (0.497–0.631)	Causes enhanced production of reactive oxygen species (ROS), which results in oxidative damage to lipids of membrane, membrane proteins, and indirect effects on DNA—indirect
UVB	280–320	3.94–4.43 (0.631–0.710)	Causes direct DNA damage by inducing the formation of DNA lesions (photoproducts), which block DNA replication and RNA transcription, oxidative stress and lipids damage, intermediate effects between UVA and UVC
UVC	100–280	4.43–12.4 (0.710–1.987)	Causes direct DNA damage by inducing the formation of DNA lesions (photoproducts), most notably pyrimidine dimers, which block DNA replication and RNA transcription

sensitivity depending on the wavelength applied. Most of the data available are for UVA LEDs that output light in the 320–400 nm range, which is the least efficient at disinfection than light in the germicidal range of UVC (200–280 nm) because it is poorly absorbed by DNA (Sinha & Häder, 2002). UVA radiation indirectly inactivates microorganisms by damaging proteins and producing hydroxyl and oxygen radicals that can destroy cell membranes and other cellular components (Sinha & Häder, 2002). This process takes more time than the damage produced by UVC, which directly affects the DNA of microorganisms. Still, Hamamoto et al. (2007) demonstrated the ability of UVA LEDs at 365 nm to inactivate bacteria in water. They found that *E. coli* DH5α was reduced by >5 log at a dose of 315 J cm^{-2} (Hamamoto et al., 2007). Also, UVA light is able to penetrate deeper into turbid liquids and could possibly offer advantages for the UV treatment of beverages.

UVB

UVB light has been known for effective treatment of various skin diseases; however, bactericidal effects of UVB LED photons have not explored widely. Takada, Matsushita, Horioka, Furuichi, and Sumi (2017) examined the bactericidal effects of UVB LEDs on oral bacteria to explore the possibility of using a 310 nm UVB LED irradiation device for treatment of oral infectious diseases. Irradiation with the 310 nm UVB LED at 105 mJ cm^{-2} showed 30%–50% bactericidal activity to oral bacteria; however, 17.1 mJ cm^{-2} irradiation with the 265 nm UVC LED completely killed the bacteria. In the same time, Ca9–22 cells in human oral epithelial cell line were strongly injured by irradiation with the 265 nm UVC LED but were not harmed by irradiation with the 310 nm UVB LED. UVB light also induced the production of ROS from oral epithelial cells and may enhance bactericidal activity to specific periodontopathic bacteria and despite a higher UV dose may be useful as a new adjunctive therapy.

Another application of UVB LEDs was explored by Argyraki et al. (2016) who tested a new type of LEDs to deliver UVB at 296 nm and UVC at 266 nm to treat *P. aeruginosa* biofilms at different growth stages. The killing rate was studied as a function of dose for 24-h-grown biofilms. The dose was ramped from 72 J m^{-2} (720 mJ cm^{-2}) to 10,000 J m^{-2} (100,000 mJ cm^{-2}). It was shown that UVB light was more effective than UVC irradiation in inactivating *P. aeruginosa* biofilms. No colony-forming units (CFU) were observed for the UVB-treated biofilms when the dose was 100,000 mJ cm^{-2} (CFU in control sample: 7.5 × 10^4). UVB irradiation at a dose of 200,000 mJ cm^{-2} on mature biofilms (72 h grown) resulted in a 3.9 log reduction efficacy. The authors concluded that the results supported the hypothesis about the importance of penetration depth, because mature biofilms create a thicker matrix expected to be less penetrable by light. More studies are needed to fully understand the mechanism of action by UVB and UVA light and what positive and negative effect they may have on certain foods and beverages.

UVC

UVC light in 200–280 nm range has previously shown to be germicidal against microorganisms such as bacteria, viruses, yeasts, molds, spores, protozoa, and algae. The extensive evidence generated shows that inactivation of bacteria, viruses, and other microorganisms occurs along the UV spectrum, but it is strongest in the UVC range between 260 and 270 nm and below 230 nm (Beck, Wright, Hargy, Larason, & Linden, 2015; Bolton, 2017). The primary mechanism of microbial inactivation is the direct absorption of UVC photons at 260 nm by DNA, which becomes damaged due to subsequent pyrimidine dimerization (Fig. 5.1), which are bonds, formed between adjacent pairs of thymine or cytosine pyrimidines on the same DNA or RNA strand. UV-induced damage to the DNA prevents it from being replicated due to damage of polymerase enzyme. The DNA replication cannot be completed and the organism cannot reproduce and infect. Although microorganisms are not killed, they have been rendered inactive. This relates for RNA-based organisms and single-stranded DNA and RNA as well.

Fig. 5.2 shows the similarity between the ability of UVC light to destroy bacterial cells and the ability of this cell's nucleic acid to absorb UVC light. Nevertheless, damage to nucleic acid does not prevent the cell from undergoing metabolism and other cell functions. Enzyme mechanisms within the cell are capable of repairing some of the damage to the nucleic acid. It is possible for microorganisms to repair themselves to the extent where they will become infective again after exposure to UV light. Microorganisms have developed two mechanisms to repair damage caused by UV light. These mechanisms are termed light and dark repair. Repair to UV light-induced DNA damage includes photoreactivation, excision repair or dark repair,

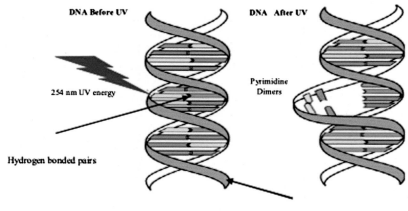

FIGURE 5.1

Structure of DNA before and after absorbing photon of UV light.

From https://en.wikipedia.org/wiki/Ultraviolet#/media/File:DNA_UV_mutation.svg.

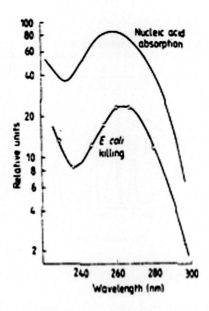

FIGURE 5.2

The absorption spectrum of DNA and inactivation *of E. coli* cells.

recombinational repair, and inducible error-prone repair. The details of the repair mechanism were reported by Jagger (1967), Shama (1992), and Harm (1980).

Fig. 5.3 illustrates the output spectra of low- and medium-pressure mercury lamps (LPM and MPM) along with the microbial action spectra. An LPM UV lamp typically converts electrical input power into resonant radiation, mostly at 254 and 185 nm. The germicidal lamp emitting UV at 254 nm is operating very close to the optimized wavelength for maximum absorption by nucleic acids. The contribution of 185 nm UV to germicidal effect is negligible, and consequently, the UV dose using LPM is a result of the near-monochromatic line at 254 nm. On the other hand, an MPM UV lamp typically generates a polychromatic spectral output over the entire UV range.

Mechanisms of inactivation can differ for some organisms due to DNA damage, damage to vital proteins, or transfer of energy in between (Beck et al., 2017). These mechanisms are consistent for both low-pressure lamps and UVC LEDs that emit light in close proximity to 260 nm.

Blue Light

Because of indirect mechanism of photodynamic microbial inactivation, the use of blue LEDs has recently received increased attention and its potential for clinical and food applications is explored individually and in combination with other wavelengths.

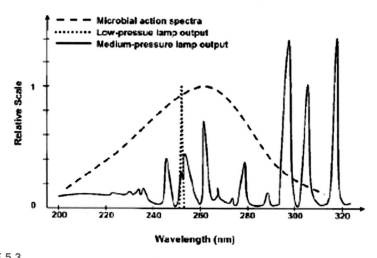

FIGURE 5.3

Output spectra of LPM and MPM lamps and microbial action spectra.

The antibacterial mechanisms of blue LED illumination could be explained by photodynamic indirect inactivation that requires photosensitizers, such as intracellular porphyrin compounds naturally produced by some bacteria, and visible lights within 400−430 nm wavelengths under the presence of oxygen (Luksienė, 2009). Once bacterial cells are exposed to light under existing oxygen, the endogenous porphyrin compounds inside cells absorb the light, followed by being excited. As a result, reactive oxygen species (ROS), such as superoxide ion and singlet oxygen, are produced. The ROS can bring about a cytotoxic effect by interacting with adjacent intracellular components, such as DNA, protein, and lipids, resulting in bacterial death (Luksienė, 2009).

A number of studies showed that blue light LEDs of 405 and 460 nm wavelengths could inactivate various foodborne pathogens, such *as E. coli* O157:H7, *L. monocytogenes*, and *S. typhimurium* in phosphate-buffered saline solution or trypticase soy broth (TSB) without the addition of exogenous photosensitizers under refrigerated condition (Ghate et al., 2013; Kim, Kim, & Kang, 2016; Kim, Mikš-Krajnik, Kumar, Ghate, & Yuk, 2015). Kumar et al. (2016) have demonstrated that 405 nm LED showed the greatest antibacterial effect compared with 460 and 520 nm LEDs. Another reported study has also shown that 405 nm wavelength revealed the greatest antibacterial efficacy on *L. monocytogenes* within the wavelengths ranging from 400 to 500 nm (Endarko, MacLean, Timoshkin, MacGregor, & Anderson, 2012).

It is important to emphasize that UV light-based solutions for food safety applications require effective light delivery to biological cellular and molecular targets. Thus, the potential applications are confined by penetration depth of light to the region of interest. UV LEDs in UVA, UVB, and UVC ranges can provide advantages over conventional sources by wider ranges of wavelengths emission, thus providing efficient strategies targeting a broad range of organisms. Additionally, combining UV, blue light, and infrared LEDs can provide another purely optical hurdle to prevent microbial growth and risk of cross-contamination.

Action Spectra for Water- and Foodborne Pathogens

An action spectrum describes the spectral effectiveness of a photobiological or photochemical process of microbial inactivation. Action spectra are presented as plots or tabled values of germicidal efficacy of an organism over a range of wavelengths. In general, the action spectrum has the shape of the absorption spectrum of the photoactive molecules catalyzing the reaction. In ideal optimized UV-based process, the effective UV sources should emit high-intensity photons in the peak absorbance wavelengths of DNA that is the germicidal target of UV photons.

Waterborne Pathogens

The sensitivity of waterborne pathogenic organisms across the UV spectrum has been studied for the validation processes that require knowledge of the photochemical properties of the pathogens of concern and the challenge microorganisms used to represent them and specifically when polychromatic UV systems are used for water disinfection. In particular, the organisms' dose-responses to UV light and their sensitivity across the UV spectrum must be known. Beck et al. (2015) measured the UV spectral sensitivity of *Cryptosporidium parvum*, *Bacillus pumilus* spores, and MS2, T1UV, Q Beta, T7, and T7m Coliphages using a tunable laser in the UVC and UVB range of 240–290 nm. These organisms are identified as the pathogens of concerns in water along with commonly used surrogates. As shown in Fig. 5.4, all tested bacteria and viruses exhibited relative peak of sensitivity between 260 and 270 nm. Of the coli phage, MS2 exhibited the highest relative sensitivity below 240 nm, relative to its sensitivity at 254 nm, followed by Q Beta, T1UV, T7m, and T7 coli phage (date are not shown).

Also, UVC LED unit at 260 nm, 280 nm, and their combinations was evaluated for microbial and viral efficacy and energy efficiency at inactivating *E. coli*, MS2 coliphage, human adenovirus type 2 (HAdV2), and B. *pumilus* spores and compared with conventional LPM and MPM lamps (Beck et al., 2017). It was found that all five UV sources demonstrated similar inactivation of *E. coli*. However, for MS2, the 260 nm UVC LED was the most effective source, whereas for HAdV2 and *B. pumilus*, the MPM UV lamp was the most effective.

Chen, Craik, and Bolton (2009) found that the action spectrum of *B. subtilis* spores deviated significantly from the relative absorbance spectrum of the DNA purified from the spores but matched quite well with the relative absorbance spectrum of decoated spores. It was concluded that the photons absorbed by components other than DNA in spores may be involved in the inactivation effect by UVC light, when

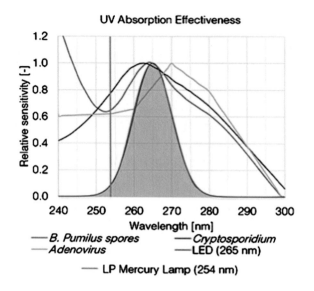

FIGURE 5.4

Wavelength sensitivity of waterborne organisms.

Adapted from Beck, S. E., Wright, H. B., Hargy, T. M., Larason, T. C., & Linden, K. G. (2015). Action spectra for validation of pathogen disinfection in medium-pressure ultraviolet (UV) systems. Water Research, 70, 27–37. https://doi.org/10.1016/j.watres.2014.11.028.

the absorbance of these components is not negligible. On the other hand, the action spectrum of *S. typhimurium* bacteria, distinctive waterborne pathogen, matched quite well with the relative absorbance spectrum of DNA extracted from vegetative cells, except in the region below 240 nm.

From food safety perspective, these reports are important because water is not only the essential ingredient of food products but also is used for washing and rinsing of raw commodities and equipment.

Foodborne Pathogens

To optimize food safety applications of UV LEDs against foodborne pathogens and other organisms, the wavelengths also should be optimized to match the target organisms and meet efficacy requirements. Green et al. (2018) measured the action spectra of three most common foodborne pathogenic strains of *E. coli, Listeria, and Salmonella* cocktail and three nonpathogenic counterparts including *E. coli* ATCC 8739, *L. innocua* ATCC 51742, and *Enterococcus faecium* using multiple-wavelength UV LEDs at 259, 268, 275, 289, and 370 nm and a single LPM lamp at 253.7 nm as a base line. Also, the bacterial spectral UV sensitivity of tested bacteria was compared with UV absorbance of their DNA and cell suspension. At 253.7 nm after exposure to equivalent UV, the dose of 7 mJ cm^{-2}, the *Salmonella* cocktail showed the highest UV sensitivity among all tested strains with logarithmic count reduction (LCR) of 4.67 ± 0.18. Among the individual pathogenic strains, *L. monocytogenes* and *S. enterica* ssp. *diarizonae* showed the highest UV sensitivity at 253.7 nm (4.03 ± 0.32 and 4.00 ± 0.28 LCR, respectively) followed by *E. coli* O157:H7 (3.24 ± 0.32 LCR).

The action spectra of six tested bacteria and *Salmonella* cocktail are shown in Fig. 5.5. At the similar equivalent UV dose of 7 mJ cm^{-2}, out of five wavelengths tested, UVC LEDs emitting at 259 and 268 nm achieved the highest log count reductions. Among the three pathogenic bacteria, *S. enterica* ssp. *diarizonae* and *E. coli*

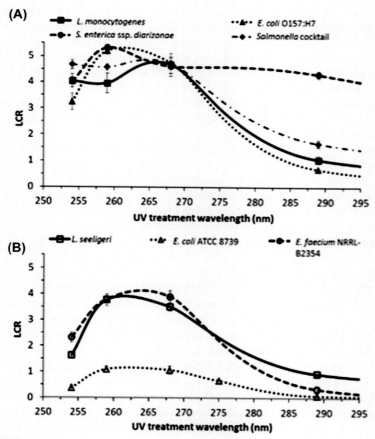

FIGURE 5.5

The UV action spectra of *L. monocytogenes*, *E. coli* O157:H7, *S. enterica* ssp. *diarizonae* (A); *Salmonella* cocktail, *L. seeligeri*, *E. coli* ATCC 8739, and *E. faecium* NRRL-B2354 (B) following UV treatment in saline suspensions at 254 nm using the LPM lamp, at 259, 268, 275, and 289 nm using UV LEDs.

O157:H7 exhibited the highest overall UV sensitivities, which occurred at 259 nm with LCR values of 5.32 ± 0.15 and 5.21 ± 0.13, respectively. In contrast, *L. monocytogenes* peaked in UV sensitivity at 268 nm, with an LCR of 4.68 ± 0.13. The *Salmonella* cocktail showed maximal and nonsignificantly different UV sensitivity between 254 and 268 nm. Also, with the exception of *S. enterica* ssp. *diarizonae*, all tested bacteria showed a progressive decrease in UV sensitivity at wavelengths beyond 268 nm. Negligible LCRs were observed following UVA treatment at 370 nm with all tested bacteria (Fig. 5.5).

All tested nonpathogenic strains showed maximal UV sensitivity at both 259 and 268 nm with no significant difference between these two wavelengths. More variability was observed among the pathogenic strains. Overall, *E. coli* ATCC 8739 was shown to be the least UV-sensitive bacterium tested in this study, showing significantly lower LCR values ($P < .05$) at all wavelengths between 259 and 289 nm. The differences in action spectra can be seen when comparing pathogenic and nonpathogenic indicator bacteria, for example, *Listeria*. The germicidal action spectrum of *Listeria seeligeri* showed a broad UV sensitivity peak between 259 and 268 nm, whereas *Listeria monocytogenes* had more narrow peak at 268 nm. Although relatively minor, these differences can be exploited by UV LEDs because they can be tuned to emit UV light at the optimal wavelength for the inactivation of a particular target organism(s).

As expected, the peak inactivation wavelengths for all tested bacteria aligned fairly well with their respective DNA absorbance profiles, which peaked at approximately 260 nm and showed a progressive decrease beyond this wavelength (Fig. 5.6). Also, a slight increase (up to 12 nm) in the spectral action peak in comparison with the DNA absorbance peak was observed in all cases. Again, the single exception to this trend was *S. enterica diarizonae*, whose action spectrum in the UVB region did not correlate to its DNA absorbance profile. Also, the UV absorbance of cell suspensions peaked at a similar wavelength range of 254–258 nm and decreased steadily as wavelength increased.

The most intriguing aspect of UV LED inactivation occurs in the far UVC, UVB, and UVA range because inactivation in these ranges can occur by means other than DNA damage. Specifically, UV light in close proximity to 280 nm (far UVC) is absorbed by aromatic amino acids and hence is believed to induce protein damage in microorganisms. Kim, Tang, Bang, and Yuk (2017) and Kim, Bang, and Yuk (2017) have shown that UV treatment of foodborne pathogens such as *E. coli* O157:H7, *Salmonella* spp., *L. monocytogenes*, and *S. aureus* at 279 nm compromises membrane integrity five- to eightfold more so than at 266 nm. Also, membrane damage at 279 nm was twofold higher in gram-positive bacteria than in gram-negative bacteria. Green et al. (2018) have compared the germicidal UV sensitivity of seven foodborne *Salmonella enterica* subspecies between 254 and 365 nm. The authors discovered that one serovar (*diarizonae* sv. 11:k:z53) showed higher UV sensitivity at 289 nm in comparison with a *Salmonella* cocktail containing all seven subspecies whose sensitivity decreased significantly at this wavelength (Fig. 5.7).

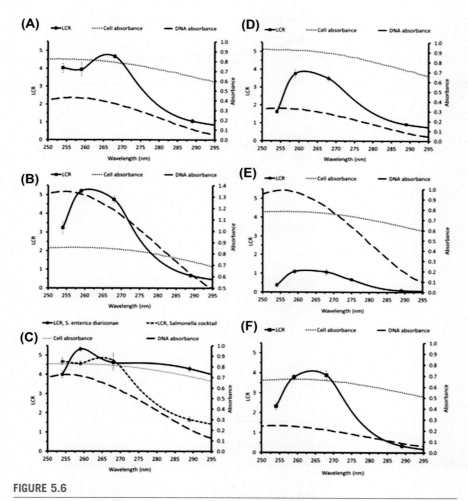

FIGURE 5.6

Action spectrum of pathogenic *L. monocytogenes* (A), *E. coli* O157:H7 (B), *S. enterica* ssp. *diarizonae* and the *Salmonella* cocktail (C) and nonpathogenic *L. seeligeri* (D), *E. coli* ATCC 8739 (E), and *E. faecium* NRRL-B2354 (F) compared with their respective DNA and cellular absorbance spectra.

This study suggested that UV inactivation of *S. diarizonae* sv 11:k:z53 occurred due to other mechanisms in addition to DNA damage.

The same authors also showed that UV inactivation kinetics of nonpathogenic *E. coli* ATCC 8739 in saline suspension differed at wavelength of 253.7 and 275 nm (Fig. 5.8).

Specifically, inactivation at 275 nm had linear character with a lack of shouldering or tailing regions and ultimately produced a slightly higher log count

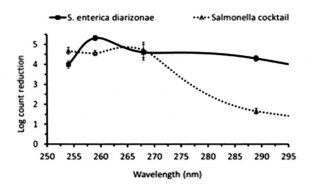

FIGURE 5.7

Log count reduction of *S. diarizonae* sv. 11:k:z53 and a seven-strain *Salmonella* cocktail following UV treatment in saline suspension at wavelengths ranging between 253.7 and 289 nm (Green et al. 2018).

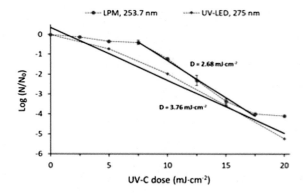

FIGURE 5.8

Inactivation kinetics of *E. coli* ATCC 8739 in saline suspension following UV treatment using an LP mercury lamp at 253.7 nm and a UV LED at 275 nm (Green et al. 2018).

reduction than at 253.7 nm. This occurred in spite of the lower observed D value (UV dose required to achieve one log count reduction) at 273.7 nm compared with 275 nm. Several other authors have also reported evidence of protein damage or enzyme inactivation using far UVC or UVB LEDs. Although not as effective as UVC and UVB LEDs, UVA LEDs have also shown to be germicidal at higher UV doses.

These reported results highlighted the notion that UV treatment of foodborne pathogens at 254 nm is not the most effective option. Given that 259 nm is closer to the bacterial DNA absorbance peak (~260 nm) than 254 nm, it is expected that this 4 nm shift in emission was responsible for the higher observed inactivation.

Multiple Wavelengths Application of UV and Visible LEDs

An important advantage of LEDs is the possibility to combine multiple wavelengths of light to improve efficacy of treatment. Several studies have shown that the use of UV LEDs emitting light at different wavelengths can lead to synergistic inactivation of microorganism as well as enzymes (i.e., a greater response (reduction) under combined use of multiple-wavelength LDE than the sum of the response after exposure to individual wavelength LDE). For example, Green et al. (2018) reported a synergistic effect during simultaneous exposure of *E. coli* ATCC 8739 to combination of 259/289 nm. The authors observed a 2.42 LCR using 259/289 nm in combination as opposed to the theoretical additive reduction of 1.2 log after using these wavelength LEDs as individual treatments. Akgun and Unluturk (2017) also showed that destruction of PPO enzyme in clear apple juice was more effective with UVC/UVA LEDs or UVC/405 nm combination than using UVC alone. Specifically, the authors observed a residual PPO activity of 56.4% using UV LED treatment at 280 nm with 771 mJ cm^{-2} in comparison with 32.6% and 34.4% with 280/365 and 280/405 nm treatments at the same UV dose, respectively. Further research is required to determine the exact cause of this phenomenon as well as to determine the food organisms or proteins that are susceptible to it because synergistic UV effects at multiple wavelengths are not commonly observed and only seem to occur with some proteins or strains of bacteria. Using a slightly different strategy through successive exposure to UVA and UVC LEDs, Xiao, Chu, He, Liu, and Hu (2018) showed that pretreatment with UVA LEDs at 365 nm increased LCR of *E. coli* ATCC 11229, 15597, and 700891 by 1.2, 1.4, and 1.2-fold, respectively, following treatment with UVC LEDs at 265 nm. However, the same authors also found that the fourth *E. coli* strain (ATCCC 25922) became more resistant to UVC treatment following UVA pretreatment. Another potential way to utilize multiple wavelengths for food treatment is to combine germicidal UV LEDs with visible and blue light LEDs. As stated earlier, visible LEDs are capable of enhancing or delaying the loss of nutritional content and promoting or delaying ripening of plant crops. Therefore, the use of UV LEDs for microbial inactivation in combination with visible LEDs can be used as a treatment strategy to delay or prevent food spoilage while enhancing the nutritional value of postharvest crops.

LEDs for Food Safety Applications

The inactivation of food pathogens and fungal spoilage organisms using UV LEDs has been reported in liquid media (ice) as well as on contact surfaces (stainless steel, cutting boards) (Table 5.2). UV LEDs were tested for treatment of sliced cheese, cabbage tissue, and raw chicken meat. To this date, the inactivation of the following bacterial food pathogens has been reported using UV LEDs: *Listeria*, *Escherichia coli*, *Salmonella*, *Campylobacter*, and *Staphylococcus*. In addition, the inactivation of two fungal spoilage organisms, *Pichia* and *Saccharomyces*, has been reported.

Table 5.2 Reported Studies on Inactivation of Food Pathogens Using UV LEDs.

Food Pathogen	Wavelength (nm)	Treatment Media	Incident Irradiance (mW cm⁻²)	UV Fluence (mJ cm⁻²)	Log Count Reduction (LCR)	References
Listeria monocytogenes	259	Saline	0.014	7.0	3.95	Green (2018)
	268		0.092		4.68	
	289		0.157		1.05	
	370		4.66		0.18	
	266	Solid media	0.0042	0.6	4.29	Kim (2017)
	271		0.0040		4.24	
	276		0.0038		3.61	
	279		0.0038		3.71	
	266	Sliced cheese surface	0.004	3.0	3.52	Kim (2016)
	271				3.94	
	276				3.31	
	279				3.24	
	270–280	Ice	0.084	160	>5	Murishita (2017)
Escherichia coli O157:H7	259	Saline	0.014	7.0	5.21	Green (2018)
	268		0.092		4.88	
	289		0.157		0.67	
	370		4.66		0.06	
	266	Solid media	0.0042	0.4	>5	Kim (2017)
	271		0.0040	0.4	>5	
	276		0.0038	0.2	>5	
	279		0.0038	0.2	>5	
	266	Sliced cheese surface	0.004	3.0	4.88	Kim (2016)
	271				4.81	
	276				4.31	
	279				4.04	
	270–280	Ice	0.084	160	<5	Murishita (2017)
Escherichia coli DH5α	365	Cabbage tissue	125.0	675,000	3.23	Aihara (2014)
	365	Orange juice	70	147,000	0.4–1.6	Lian (2010)
	365	PBS	70	147,000	2.5	
	365	Water	70	315,000	5.7	Hamamoto (2007)
Escherichia coli EPEC	365	Water	70	252,000	>5.2	Hamamoto (2007)

Organism	nm					Reference
Salmonella cocktail	259	Saline	0.014	7.0	4.57	Green (2018)
	268		0.092		4.72	
	289		0.157		1.65	
Salmonella spp.	266	Solid media	0.0042	0.4	>5	Kim (2017)
	271		0.0040	0.4	>5	
	276		0.0038	0.4	>5	
	279		0.0038	0.6	>5	
Salmonella enterica sv. Typhimurium	266	Sliced cheese surface	0.004	3.0	4.72	Kim (2016)
	271				4.73	
	276				4.24	
	279				3.91	
Salmonella enteritidis	270–280	Ice	0.084	160	>5	Murishita (2017)
	365	Water	70	504,000	3.4	Hamamoto (2007)
Staphylococcus aureus	266	Solid media	0.0042	0.6	4.71	Kim (2017)
	271		0.0040		4.49	
	276		0.0038		4.45	
	279		0.0038		3.84	
	365	Water	70	252,000	>5.2	Hamamoto (2007)
Pichia membranaefaciens	266	Solid media	0.0042	0.6	4.53	Kim (2017)
	271		0.0040		3.77	
	276		0.0038		4.44	
	279		0.0038		3.19	
Saccharomyces pastorianus	266	Solid media	0.0042	0.6	1.62	Kim (2017)
	271		0.0040		1.43	
	276		0.0038		1.2	
	279		0.0038		0.78	
Campylobacter jejuni	395	MRD solution	30	9000	<6	Haughton (2012)
		Raw skinless chicken fillet	30	1800	2.62	
		Chicken skin	30	1800	0.57	
		Stainless steel	30	1800	>2	
		Cutting board	30	1800	>4	
Vibrio parahaemolyticus		Water	70	252,000	>5.2	Hamamoto (2007)

The most UV-resistant food-related organism tested was *Pichia membranaefaciens*—common food spoilage yeast.

Bacterial foodborne pathogens were typically more UV sensitive than spoilage organisms and were successfully inactivated by 3.8—5 logs in clear solution or on contact smooth surfaces using LEDs in UVC range (259—280 nm), with one study in the UVB range (289 nm) and three studies in the UVA range (365, 370, and 395 nm) (Table 5.2). The LCR of bacterial or fungal food pathogens was dependent on UV wavelength, microbial strain as well as the treatment media. Studies involving UV LEDs using relatively clear solutions (saline) or smooth surfaces (solid growth media) typically required lower UV doses between 0.2 and 7.0 mJ cm^{-2}. Treatments using UVC LEDs had higher germicidal efficiency (greater inactivation per UV dose) when compared with UVB and UVA LEDs. Nonetheless, Haughton et al. (2012) observed a notable reduction (>4 log) of *C. jejuni* on a cutting board surface at 395 nm with a UV dose of 1800 mJ cm^{-2}. Sliced cheese was the only food surface treated using UVC light. Surprisingly, both *L. monocytogenes* and *S. typhimurium* were reduced up to 3.94 and 4.73 log, respectively, at 271 nm and UV dose of only 3.0 mJ cm^{-2} (Kim et al., 2016) UVA LEDs were used to treat both *E. coli* DH5α on cabbage tissue, in orange juice (365 nm) and *C. jejuni* on raw skinless chicken fillets and raw chicken skin (395 nm). Two studies reported notable log reductions of 3.23 and 2.62 on cabbage tissue and raw skinless chicken fillets using UV doses of 675 and 1.8 J cm^{-2}, respectively. Hamamoto, Mori, Takahashi et al. (2007) also reported 0.4 and 1.6 log reduction values of *E. coli* DH5α in two different commercial orange juices using UVA LED treatment at 365 nm with 147 J cm^{-2}. However, the juice samples were not stirred during treatment, which could have hindered inactivation. Germicidal effects in the UVA range are worth further perusing as this LED treatment option can be favorable because of reduced cost and increased optical power output in comparison with UVC light. Finally, the inactivation of fungal food organisms such as *Penicillium*, *Aspergillus*, *Fusarium*, and yeasts, particularly on food surfaces, using UV LEDs, is another area that requires attention, as fungal growth is the primary cause of fresh produce spoilage.

There have been comparatively fewer studies involving UV LEDs in the UVC range, likely because of their higher cost and comparatively lower irradiance. Currently, complete action spectra for many common foodborne pathogens are lacking and such information could allow for the selection of UV wavelengths for disinfection based on the peak sensitivity of target organisms, increasing efficacy over LPM lamps. However, the small size of LEDs lends itself well to a variety of applications. Until optical power output values reach that of LPM lamps, the industrial food applications of UVC and UVB LEDs will likely involve low irradiance long-term exposure or point-of-use applications. As stated earlier, the extension of operational lifetimes by means of intermittent or pulsed use enables UV LEDs to be utilized for long-term exposure.

UV LEDs for Fresh Produce

UVC light has been shown to be effective against pathogens and spoilage organisms on the surfaces of a diverse selection of produce. These include broccoli, leafy green produce, mangoes, and many other fruits and vegetables. However, there have been comparatively fewer studies involving UV LEDs likely because of their higher cost, comparatively lower power output, and lack of LEDs units.

Our research group at Agriculture and Agri-Food Canada (AAFC) is currently comparing the inactivation of fungal and bacterial pathogens on the surface of apples and leafy green following UV treatment with either an LPM lamp or UVC LEDs at 277 nm that has been determined as the optimal wavelength(s) for use against food-borne pathogens *E. coli*, *L. monocytogenes*, and *S. enterica*. 277 nm UVC LEDs were chosen above other wavelengths as it was the best trade-off between photon penetration, germicidal efficacy, and cost. The custom-made PearlBeam unit (Aqui-Sense, KY) consisted of 18 UVC LEDs (277 nm) arranged on two circular chips (9 LEDs each) mounted on the top of a plastic box with an open bottom (Fig. 5.9).

Fresh apples and romaine lettuce inoculated with *E. coli* O157:H7 and *Listeria* monocytogenes were treated using single LPM lamp at 253.7 nm and UVC LEDs at 277 nm. In addition, the effect of 277 nm was compared against pathogenic *Listeria* and spoilage spores of *Penicillum expansum*. It has been observed that treatment of *L. monocytogenes* and *P. expansum* on apple surface using an LPM lamp and UVC LEDs at 277 nm resulted in similar log inactivation values of 1.5 and 3.0 logs, respectively. However, UV LEDs were the more efficient option as they were capable of achieving this inactivation with the lower UV dose. The opposite trend was observed with *E. coli* O157:H7 on the surface of romaine as treatment with an LPM lamp was more efficient than the UVC LEDs. It was also found that to achieve similar inactivation of approximately 1.2−1.3 logs of *Listeria monocytogenes* on the surface of apples and lettuce, much higher dose of 3000 mJ cm^{-2} was needed for apple treatment compared with 100 mJ cm^{-2} for lettuce (Fig. 5.10).

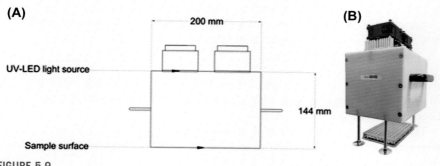

FIGURE 5.9

Schematic diagram (A) and photo of the custom UV LED unit (B) (Aquisense, KY).

Used with permission from AquiSense Technologies, KY, USA.

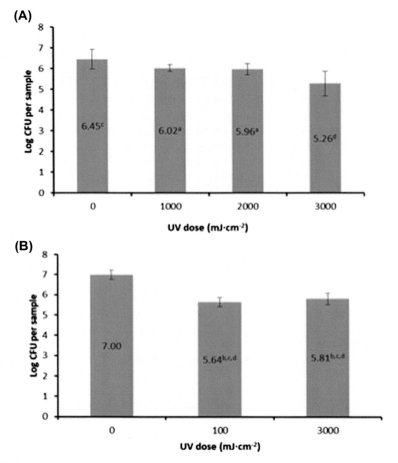

FIGURE 5.10

Log CFU recovered from apple (A) and lettuce (B) samples inoculated with
L. monocytogenes and treated with varying doses of UV light ([a,b,c,d]). Averages with the
same superscript letter are not significantly different.

Also, 277 nm UV LEDs showed an increase in irradiance from 2.33 to
2.41 mW cm^{-2} at the sample surface when operated at refrigerated temperature
(Green et al., 2018).

These data showed the potential of using LEDs in UVC range as an emerging
intervention to control pathogens on fresh produce and food contact surfaces in storage and processing areas.

Blue and Visible Light LEDs

LEDs emitting at wavelengths in near-UV-visible range have been shown by a number of researchers to be extremely effective in reducing microbial loads in a variety of foods (Table 5.3) (Kim, Bang et al., 2017; Kim, Tang et al., 2017; Lacombe et al., 2016; Srimagal, Ramesh, & Sahu, 2016). Kim, Tang, Bang, and Yuk (2017) and Kim, Bang, and Yuk (2017) have shown that blue light LEDs at wavelengths of 405 nm are able to reduce levels of *Salmonella* on fresh-cut papaya by up to 1.3 LCR at a dose of $1.7\,\text{kJ cm}^{-2}$. Fresh-cut mango was exposed to doses up to $3.5\,\text{kJ cm}^{-2}$ to study the effect on three food pathogens. The reduction up to 1.6 LCR was found, and dependency on operation temperature was reported (Kim, Bang et al., 2017; Kim, Tang et al., 2017). Among other applications of blue light LEDs, the treatment of chicken skin, salmon, milk, and contact surfaces was investigated. To achieve microbial reduction higher than 1 log, the long time exposure was required in the range of doses between 0.2 and $3\,\text{kJ cm}^{-2}$. In addition, LEDs emitting at 405 nm have shown to be effective against *E. coli*, *Salmonella*, *Shigella*, *Listeria*, and *Mycobacterium* on agar surfaces and in saline suspension, resulting in reductions up to 5 log at UV doses up to $0.288\,\text{kJ cm}^{-2}$ by Murdoch, MacLean, Endarko, MacGregor, and Anderson (2012). Kumar et al. (2016) have examined the efficacy of LEDs at 405, 460, and 520 nm against *Lactobacillus plantarum*, *Staphylococcus aureus,* and *Vibrio parahaemolyticus*. The 405 and 460 nm LEDs were shown to be effective at high doses against these organisms, achieving up to a 4 log reduction of *V. parahaemolyticus*, while 520 nm LEDs were shown to not achieve a significant log reduction of any organism (Kumar et al., 2016).

LEDs' ability to operate with increased power output (up to 20% increased irradiance at 0 vs. 37°C) at cold temperatures (Shin, Kim, Kim, & Kang, 2015) makes them an attractive option for use in the processing and storage of fresh produce, which is usually performed at low temperatures.

UV LEDs can be integrated into commercial or domestic refrigerators as a means to prevent food spoilage and by extending shelf life to reduce waste in households. Domestic refrigerators with blue light LEDs (405 nm) are commercially available for this purpose. However, Gunther et al. (2016) and Sommers et al. (2016) have shown that the effectiveness of blue light for pathogens inactivation on poultry products under refrigeration conditions is hindered and partly due to secondary thermal effects. The implementation of UV LEDs for this purpose could be a more effective approach. Britz, Gaska, Shturm et al. (2013) have shown that low irradiance UV LED treatment (272, 289, or 293 nm) of strawberries under refrigeration conditions was successful in delaying fungal spoilage by approximately twofold following a 9-day storage period.

Effect of UV LEDs on Food Quality and Nutritional Parameters

The effect of UVC light emitted from low-pressure lamps on the quality of food and beverages post-UV treatment has been previously explored (Koutchma, Forney, &

Table 5.3 Summary of Reported Food Applications of Blue Light LEDs for Microbial Inactivation.

Organisms	Wavelength, nm Dose, kJ cm^{-2}	Food	Log Reduction Effect	References
Salmonella spp.	405 nm 1.3–1.7 kJ cm^{-2} 36–48 h, 4°C	Fresh-cut papaya	1.3	Kim et al. (2017)
Escherichia coli O157:H7 *Listeria monocytogenes Salmonella*	405 nm, 2.6–3.5 kJ cm^{-2}, 36–48 h, 4 and 10°C	Fresh-cut mango	1.0–1.6	Kim et al. (2017)
Escherichia coli Staphylococcus aureus	0.013 kJ cm^{-2}	Photosensitizer Curcumin	5.94 5.91	Bhavya et al. (2019)
Escherichia coli	405–460 nm (10 W) 406 nm, 3.8°C, and 37.8 min	Milk	>5	Srimagal et al. (2016)
Cocktails of *Salmonella* spp., *Escherichia coli, Staphylococcus* spp., *or Listeria monocytogenes*	405 nm at 0.18 kJ cm^{-2}	Chicken skin Chicken purge	0.4 0.23–0.7	Sommers et al. (2017)
Campylobacter jejuni C. coli	405 nm 0.18 kJ cm^{-2}	Chicken skin	1.7 2.1	Gunther et al. (2016)
Listeria monocytogenes biofilm	405 nm 0.8 kJ cm^{-2}	Stainless steel acrylic coupons	Biofilm growth inhibition	Li et al. (2010)
Listeria monocytogenes cocktail	460 nm and riboflavin 2.4 kJ cm^{-2}	Smoked salmon	0.7–1.2	Josewin et al. (2018)

Moraru, 2009). However, UV LED studies involving postharvest treatments of food and beverages are only beginning to emerge. For this reason, limited information is available on the effects of UV LEDs (particularly UVC and UVB LEDs) on food quality and nutritional parameters post-UV treatment. However, along with microbial inactivation, other food attributes are equally as important for assessing the feasibility of UV LED treatments in food safety and other applications. This is particularly true for lower energy wavelengths, such as far UVC, UVB, and UVA, which have been shown to inactivate food pathogens by non-DNA-related mechanisms. As with visible LEDs, UVA LEDs have also been shown to enhance the nutritional parameters of postharvest food products. Kanazawa, Hashimoto, Yoshida, Sungwon, and Fukuda (2012) reported that UVA LED treatment at 375 nm for 160 min per day for 3 days was successful in stimulating the production of flavonoids and phenylpropanoids in watercress and garden pea sprouts. Britz et al. (2013) have also shown that UV LED treatment of strawberries at 272, 289, and 293 nm preserved anthocyanin and total soluble sugar content following 9 days as opposed to nontreated samples. These results support the notion that exploring individual and combined effects on specific foods and microorganisms is important for identifying promising UV LED applications.

Conclusions

Ultraviolet (UV) LEDs have shown tremendous growth and potential for food safety applications in the last number of years. The recent technological advances in manufacturing of UVC LEDs have expedited this trend. The inactivation of food pathogens by UVC, UVB, and, in some instances, UVA LEDs has been reported in model solutions, on smooth surfaces, and a limited number of food products. To maximize many benefits of UV LEDs for food safety applications, further research is required to fully characterize the susceptibility of both food pathogens and food spoilage organisms at various wavelengths in the UVC, UVB, and UVA range. Furthermore, possible synergistic effects for each target organism should be explored to determine if single or multiple wavelengths (including visible light) are more effective for treatment. For the purposes of food safety and process development, it is critical to generate new data using food products and beverages (juices, dairy, fresh produce, meat, etc.) as certain wavelengths (or combinations) might differ in germicidal efficacy when the treatment medium is altered. It is also important to evaluate the effects of various wavelengths on food quality and nutritional parameters such as color, enzyme activity, antioxidants, total phenols, vitamins, and food sensory attributes.

Likewise, UV LEDs can be first implemented in industrial food storage or transportation facilities for the treatment of fresh produce, as well as food contact surfaces (cutting boards, cutlery, stainless steel, etc.) and food packaging. The LEDs small size allows for improved positioning options and easy integration into processing units such as continuous flow reactors, which allows for better control of hydrodynamics and ultimately more flexibility in terms of reactor design. As more testing and validation work is conducted, scalable, tunable, configurable, and tailored multiple-wavelength LED-based light modules will emerge as environmentally friendly solutions to save energy, water, reduce costs, lower reliance on toxic chemicals, improve worker and consumer safety, and extend fruit and vegetable shelf life. UV LEDs present a new technological solution with enormous potential for control of pathogens and spoilage throughout the supply chain from farm to fork.

References

Aihara, M., Lian, X., Shimohata, T., Uebanso, T., Mawatari, K., Harada, Y., Akutagawa, M., Kinouchi, Y., & Takahashi, A. (2014). Vegetable surface sterilization system using UVA light-emitting diodes. *The journal of medical investigation, 61*(3-4), 285−290.

Akgun, M. P., & Unluturk, S. (2017). Effects of ultraviolet light emitting diodes (LEDs) on microbial and enzyme inactivation of apple juice. *International Journal of Food Microbiology, 260,* 65−74.

Argyraki, A., Markvart, M., Nielsen, A., Bjarnsholt, T., Bjørndal, L., & Petersen, P. M. (2016). Comparison of UVB and UVC irradiation disinfection efficacies on *Pseudomonas aeruginosa* biofilm. *Proceedings of SPIE, 988730.* https://doi.org/10.1117/12.2225597. SPIE − International Society for Optical Engineering.

Beck, S. E., Ryu, H., Boczek, L. A., Cashdollar, J. L., Jeanis, K. M., Rosenblum, J. S., et al. (2017). Evaluating UV-C LED disinfection performance and investigating potential dual-wavelength synergy. *Water Research, 109*(Suppl. C), 207−216. https://doi.org/10.1016/j.watres.2016.11.024.

Beck, S. E., Wright, H. B., Hargy, T. M., Larason, T. C., & Linden, K. G. (2015). Action spectra for validation of pathogen disinfection in medium-pressure ultraviolet (UV) systems. *Water Research, 70,* 27−37. https://doi.org/10.1016/j.watres.2014.11.028.

Bhavya, M., & Hebbar, H. U. (2019). Efficacy of blue LED in microbial inactivation: Effect of photosensitization and process parameters. *International Journal of Food Microbiology, 290,* 296−304.

Bolton, J. R. (2017). Action spectra: A review. *IUVA News, 19*(2).

Britz, S., Gaska, I., Shturm, I., Bilenko, Y., Shatalov, M., & Gaska, R. (2013). *Deep ultraviolet (DUV) light-emitting diodes (LEDs) to maintain freshness and phytochemical composition during postharvest storage. CLEO: 2013; 2013 2013/06/09.* San Jose, California: Optical Society of America.

Chen, R. Z., Craik, S. A., & Bolton, J. R. (2009). Comparison of the action spectra and relative DNA absorbance spectra of microorganisms: Information important for the determination of germicidal fluence (UV dose) in an ultraviolet disinfection of water. *Water Research, 43*(20), 5087−5096. https://doi.org/10.1016/j.watres.2009.08.032.

EFSA. (2016). *Safety of UV-treated milk as a novel food pursuant to regulation (EC) No. 258/97.* EFSA. http://www.efsa.europa.eu/en/efsajournal/pub/4370.

Endarko, E., MacLean, M., Timoshkin, I., MacGregor, S., & Anderson, J. (2012). Highintensity 405-nm light inactivation of Listeria monocytogenes. *Photochemistry and Photobiology, 88,* 1280−1286.

Ghate, V. S., Ng, K. S., Zhou, W., Yang, H., Khoo, G. H., Yoon, W.-B., et al. (2013). Antibacterial effect of light emitting diodes of visible wavelengths on selected foodborne pathogens at different illumination temperatures. *International Journal of Food Microbiology, 166,* 399−406.

Green, A. (2018). *UV light-based intervention to inactivate pathogens on fresh produce through use of UV-LEDs and water-assisted processing* (M.Sc thesis presented to the University of Guelph, December 2018).

Green, A., Popović, V., Pierscianowski, J., Biancaniello, M., Warriner, K., & Koutchma, T. (2018). Inactivation *of Escherichia coli, Listeria* and *Salmonella* by single and multiple wavelength ultraviolet-light emitting diodes. *Innovative Food Science & Emerging Technologies, 47,* 353−361.

Gunther, N. W., Phillips, J. G., & Sommers, C. (2016). The effects of 405-nm visible light on the survival of *Campylobacter* on chicken skin and stainless steel. *Foodborne Pathogens and Disease, 13*(5), 245–250.

Hamamoto, A., Mori, M., Takahashi, A., Nakano, M., Wakikawa, N., Akutagawa, M., et al. (2007). New water disinfection system using UVA light-emitting diodes. *Journal of Applied Microbiology, 103*, 2291–2298.

Harm, W. (1980). *Biological effects of ultraviolet radiation*. Cambridge University Press.

Haughton, P. N., Grau, E. G., Lyng, J., Cronin, D., Fanning, S., & Whyte, P. (2012). Susceptibility of *Campylobacter* to high intensity near ultraviolet/visible 395+/−5 nm light and its effectiveness for the decontamination of raw chicken and contact surfaces. *International Journal of Food Microbiology, 159*(3), 267–273.

Jagger, J. (1967). *Introduction to research in UV photobiology*. Englewood Cliffs, N.J: Prentice-Hall, Inc.

Josewin, S. W., Ghate, V., Kim, M.-J., & Yuk, H.-G. (2018). Antibacterial effect of 460 nm light-emitting diode in combination with riboflavin against *Listeria monocytogenes* on smoked salmon. *Food Control, 84*, 354–361.

Kanazawa, K., Hashimoto, T., Yoshida, S., Sungwon, P., & Fukuda, S. (2012). Short photoirradiation induces flavonoid synthesis and increases its production in postharvest vegetables. *Journal of Agricultural and Food Chemistry, 60*(17), 4359–4368.

Kim, M. J., Bang, W. S., & Yuk, H. G. (2017). 405 +/− 5 nm light emitting diode illumination causes photodynamic inactivation of *Salmonella* spp. on fresh-cut papaya without deterioration. *Food Microbiology, 62*, 124–132.

Kim, M.-J., Mikš-Krajnik, M., Kumar, A., Ghate, V., & Yuk, H.-G. (2015). Antibacterial effect and mechanism of high-intensity 405±5nm light emitting diode *on Bacillus cereus, Listeria monocytogenes, and Staphylococcus aureus* under refrigerated condition. *Journal of Photochemistry and Photobiology B: Biology, 153*(3), 3–9.

Kim, M. J., Tang, C. H., Bang, W. S., & Yuk, H. G. (2017). Antibacterial effect of 405+/−5 nm light emitting diode illumination against *Escherichia coli* O157:H7, *Listeria monocytogenes*, and *Salmonella* on the surface of fresh-cut mango and its influence on fruit quality. *International Journal of Food Microbiology, 244*, 82–89.

Kim, S.-J., Kim, D.-K., & Kang, D.-H. (2016). Using UVC light-emitting diodes at wavelengths of 266 to 279 nanometers to inactivate foodborne pathogens and pasteurize sliced cheese. *Applied and Environmental Microbiology, 82*(1), 11.

Koutchma, T. N., Forney, L. J., & Moraru, C. I. (2009). *Ultraviolet light in food technology*. Boca Raton, FL: CRC Press.

Kumar, A., Ghate, V., Kim, M. J., Zhou, W., Khoo, G. H., & Yuk, H. G. (2016). Antibacterial efficacy of 405, 460 and 520 nm light emitting diodes on *Lactobacillus plantarum, Staphylococcus aureus* and *Vibrio parahaemolyticus*. *Journal of Applied Microbiology, 120*(1), 49–56. https://doi.org/10.1111/jam.12975.

Lacombe, A., Niemira, B. A., Sites, J., Boyd, G., Gurtler, J. B., Tyrell, B., et al. (2016). Reduction of bacterial pathogens and potential surrogates on the surface of almonds using high-intensity 405-nanometer light. *Journal of Food Protection, 79*(11), 1840–1845. https://doi.org/10.4315/0362-028X.JFP-15-418.

Li, J., Hirota, K., Yumoto, H., Matsuo, T., Miyake, Y., & Ichikawa, T. (2010). Enhanced germicidal effects of pulsed UV-LED irradiation on biofilms. *Journal of Applied Microbiology, 109*(6), 2183–2190. https://doi.org/10.1111/j.1365-2672.2010.04850.x.

Lian, X., Tetsutani, K., Katayama, M., Nakano, M., Mawatari, K., Harada, N., Hamamoto, A., Yamato, M., Akutagawa, M., & Kinouchi, Y. (2010). A new colored beverage disinfection system using UV-A light-emitting diodes. *Biocontrol Sci, 15*, 33–37.

Luksiene, Z. (2009). Photosensitization for food safety. *Chemine Technologija, 4*(53), 62–65.

Murdoch, L. E., MacLean, M., Endarko, E., MacGregor, S. J., & Anderson, J. G. (2012). Bactericidal effects of 405nm light exposure demonstrated by inactivation of *Escherichia, Salmonella, Shigella, Listeria*, and *Mycobacterium* species in liquid suspensions and on exposed surfaces. *The Scientific World Journal.* https://doi.org/10.1100/2012/137805.

Murashita, S., Kawamura, S., & Koseki, S. (2017). Inactivation of Nonpathogenic Escherichia coli, Escherichia coli O157:H7, Salmonella enterica Typhimurium, and Listeria monocytogenes in Ice Using a UVC Light-Emitting Diode. *Journal of Food Protection, 80*(7), 1198–1203.

Nelson, K. L., Boehm, A. B., Davies-Colley, R. J., Dodd, M. C., Kohn, T., Linden, K. G., et al. (2018). Sunlight-mediated inactivation of health-relevant microorganisms in water: A review of mechanisms and modeling approaches. *Environmental Sciences: Processes Impacts, 20*, 1089–1122.

Santos, A. L., Oliveira, V., Baptista, I., Henriques, I., Gomes, N. C. M., Almeida, A., et al. (2013). Wavelength dependence of biological damage induced by UV radiation on bacteria. *Archives of Microbiology, 195*, 63–74.

Shama, G. (1992). Ultraviolet irradiation apparatus for disinfecting liquids of high ultraviolet absorptivities. *Letters in Applied Microbiology, 15*, 69–72.

Shin, J.-Y., Kim, S.-J., Kim, D.-K., & Kang, D.-H. (2015). Evaluation of the fundamental characteristics of deep UV-LEDs and application to control foodborne pathogens. *Applied and Environmental Microbiology, 59.* https://doi.org/10.1128/aem.01186-15.

Sinha, R. P., & Häder, D. P. (2002). UV-induced DNA damage and repair: A review. *Photochemical and Photobiological Sciences, 1*, 225–236.

Sommers, C., Gunther, N. W., & Sheen, S. (2017). Inactivation of *Salmonella* spp., pathogenic *Escherichia coli, Staphylococcus spp.*, or *Listeria monocytogenes* in chicken purge or skin using a 405-nm LED array. *Food Microbiology, 64*, 135–138.

Srimagal, A., Ramesh, T., & Sahu, J. K. (2016). Effect of light emitting diode treatment on inactivation of *Escherichia coli* in milk. *Lebensmittel-Wissenschaft und -Technologie-Food Science and Technology, 71*, 378–385. https://doi.org/10.1016/j.lwt.2016.04.028.

Takada, A., Matsushita, K., Horioka, S., Furuichi, Y., & Sumi, Y. (2017). Bactericidal effects of 310 nm ultraviolet light-emitting diode irradiation on oral bacteria. *BMC Oral Health, 17*, 96. https://doi.org/10.1186/s12903-017-0382-5.

USEPA 815-R-06-007, 2006. UV guidelines (http://www.epa.gov/safewater/disinfection/lt2/pdfs/guide_lt2_uvguidance.pdf) 3rd edition.

U. S. Food and Drug Administration. (2001). 21 CFR Part 179. Irradiation in the production, processing and handling of food. *Federal Register, 65*, 71056–71058.

U.S. Food and Drug Administration. (August 29, 2012). 21 CFR part 172. § 172.381 Vitamin D2 bakers yeast. *Federal Register, 77*(168). Rules and Regulations 52229. Food additives permitted for direct addition to food for human consumption; vitamin D2 bakers yeast.

Xiao, Y., Chu, X. N., He, M., Liu, X. C., & Hu, J. Y. (2018). Impact of UVA pre-radiation on UVC disinfection performance: Inactivation, repair and mechanism study. *Water Research, 141*, 279–288.

Further Reading

Beck, S. (2017). UV LED disinfection 101. *IUVA News*.

Luksiene, Z. (2003). Photodynamic therapy: Mechanism of action and ways to improve the efficiency of treatment. *Medicina, 39*(12), 1137—1150.

Luksienė, Z., & Zukauskas, A. (2009). Prospects of photosensitization in control of pathogenic and harmful micro-organisms. *Journal of Applied Microbiology, 107*(5), 1415—1424.

Maclean, M., MacGregor, S. J., Anderson, J. G., & Woolsey, G. (2009). Inactivation of bacterial pathogens following exposure to light from a 405-nanometer light-emitting diode array. *Applied and Environmental Microbiology, 75*(7), 1932—1937.

Murdoch, L. E., Maclean, M., MacGregor, S. J., & Anderson, J. G. (2010). Inactivation of *Campylobacter jejuni* by exposure to high-intensity 405-nm visible light. *Foodborne Pathogens and Disease, 7*(10), 1211—1217.

The Nobel Foundation. (1903). *The nobel prize in physiology or medicine 1903*. Nobel Prizes and Laureates. Retrieved June 14, 2018, from: https://www.nobelprize.org/nobel_prizes/medicine/laureates/1903/.

Future Outlook

The light-mediated techniques through reactive oxidative species (ROS) production in photodynamic inactivation (PDI), photocatalytic oxidation, and direct microbial UV inactivation using blue light and UV LEDs are still maturing fields related to food production and safety. Currently, the above mechanisms of microbial inactivation are mainly researched and more established for decontamination in medical applications, water and air disinfection. However, focus is now shifting to the applicability of light in food-related decontamination and inactivation processes, with LEDs having a major role as a suitable source of UV energy. The reported studies in vitro have shown the effectiveness of UV LEDs directly inactivating a variety of significant foodborne and waterborne pathogens at the multiple wavelengths. More research is needed to be conducted on actual food matrices and food products. Photosensitizers and photocatalytic materials can also be incorporated into food coatings or packaging or contact surfaces materials so that efficient indirect inactivation of such surfaces can be achieved using LEDs. With further progress in LED technology, LED systems can be further tested for postharvest operations to extend storage and in maintaining sanitation and safety of raw and finished foods and food facilities. In addition to the established applications in UV curing, printing and medical areas, and despite technology barriers at shorter wavelengths, there is a clear potential for LEDs market penetration in produce production, postharvest, and food safety industries. UVC LEDs integration into domestic and commercial refrigerators for disinfection purposes also has huge potential.

Schematic diagram on Fig. A1.1 shows existing and projected applications across food production areas for visible and UV LEDs.

The cost of UV LEDs has dropped significantly during the past several years through improvements in manufacturing technologies. However, when compared with the price of visible white LEDs, UV prices are still much higher and fewer manufacturers supply UV LEDs—although volumes are high for these suppliers. Overall, LEDs have come a long way from invention to practical applications, and their practicality in the food industry is becoming increasingly evident. LED technology can bring unique benefits to the food supply chain, from the production of produce, to the postharvest stage, and during processing to provide food safety before human consumption in the kitchens. The further development of LEDs will be of great benefit to the food production industry and consumers.

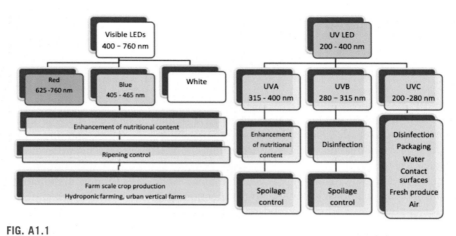

FIG. A1.1

Established and potential applications of LEDs in food production and processing and safety.

Glossary of Common Terms of UV LEDs and Food Applications

Artificial lighting or grow lights An electric light designed to stimulate plant growth by emitting a light appropriate for photosynthesis. Artificial lighting is used in applications where there is either no naturally occurring light or where supplemental light is required as in the winter months. Artificial lighting attempts either to provide a light spectrum similar to that of the sun or to provide a spectrum that is more tailored to the needs of the plants being cultivated.

Chip A very small square of semiconducting material also known as a "die" it is the "active" light-emitting component of LEDs.

Diode A semiconductor device that conducts electric current in one direction only.

Disinfection Process that eliminates many or all pathogenic microorganisms, except bacterial spores, on inanimate objects through the use of chemicals, heat, or physical agents.

Efficiency Wall plug efficiency (or radiant efficiency, WPE) is the ratio of optical power output to input electrical power.
Germicidal (or UVC) efficiency (GE) is the ratio of the output optical power output at germicidal range (often at 253.7 nm) to input electrical power.

Exposure Effective radiant energy density at a surface; the time integral of irradiance (fluence rate) within a specified bandwidth expressed in $J\,m^{-2}$ or $mJ\,cm^{-2}$.

Food preservation Procedures and processing operations targeted toward pathogenic and spoilage microbiota in food using chemical or physical methods that are relying either on the inhibition of microbial growth or on microbial inactivation.

Food safety A scientific area describing handling, preparation, preservation, and storage of food in ways that prevent foodborne illness. Food safety also refers to the conditions and practices that preserve the quality of **food** to prevent contamination by **food**borne organisms.

Foodborne pathogens Substances such as parasites, viruses or bacteria in food that cause illness or food poisoning, and in severe cases, death. Although every outbreak of foodborne illness is different, the key foodborne pathogens are *Escherichia coli, Salmonella, Listeria, Cyclospora,* hepatitis A, norovirus.

Fluence The total radiant energy of all wavelengths passing from all directions through infinitely small sphere of cross-area dA divided by A, measured in $mJ\,cm^{-2}$ or $J\,m^{-2}$.

Horticulture The area of agriculture that deals with the art, science, technology, and business of growing plants. It includes the cultivation of medicinal plants, fruits, vegetables, nuts, seeds, herbs, sprouts, mushrooms, algae, flowers, seaweeds, and nonfood crops such as grass and ornamental trees and plants.

Intensity Not well-defined generic term sometimes means irradiance expressed in $mW\,cm^{-2}$ or $W\,m^{-2}$.

Irradiance Radiant energy arriving at the surface from all directions per unit area expressed in $mW\,cm^{-2}$ or $W\,m^{-2}$.

Lead frame A metallic frame used for mounting and connecting LED chips. The lead frame functions as the electrical path for the device.

Lifetime The time in hours required for UV light output of the source to reach 80%–85% of its original value.

Light-emitting diodes (LEDs) A semiconductor diode or chip capable of producing light through electroluminescence and convert electrical energy into visible, infrared, or ultraviolet light depending on the semiconductor material.
Infrared LEDs: Special-purpose LED that transmits IR rays in the range of wavelength from 760 up to 3000 nm.
Visible LEDs: Blue, red, and white light chips that emit light in the range from 406 to 760 nm.
UV LEDs: They include UVA, UVB, and UVC LEDs and can be manufactured to emit light in the ultraviolet range from 250 to 400 nm.

LEDs chip A single diode that is made through metal vapor phase deposition process using special wafers. Diode or chip is composed of two semiconductor layers—an n-type layer that provides electrons and a p-type layer that provides holes for the electrons to fall into.

LED package Multiple chips placed in a protective package an assembly, electrically assembled together and arranged in a matrix pattern with optics and solderable leads for optimum performance.

LED modules Packaged assembly consisting of one or multiple LED diodes that are individually wire bonded to a printed circuit board, which is then secured to a heat sink; LED packages joined together to form large LED array.

LED system Integrated bare chips or packaged chips built in a single fixture to use defined wavelength and output power for specific applications.

Lethal agents Physical and chemical agents affecting microbial activities to such an extent as depriving microbial particles of the expected reproductive capacity. Heat, ionizing, and UV light are the most relevant physical lethal agents.

Logarithmic reduction or log reduction (LR) Achieved by a disinfection process is a measure of the reduction in the concentration of contaminated organisms. A logarithmic scale with a base 10 is used to evaluate the reduction of microbial concentration after applying lethal antimicrobial agents. Log reduction stands for a 10-fold (or one decimal point) reduction in bacteria, meaning the lethal agent reduces the number of live bacteria by 90%. There are a few ways of describing a reduction of 100 to 10: a reduction of 1 log; a 90% reduction; a reduction by a factor of 10; a reduction to 1/10 of the original value; the result is 10 times smaller than the original value.

Mercury lamps Special lamps that produce UVC light via the vaporization and ionization of mercury.

Microbial inactivation Process of using physical or chemical agents that cause microbial cell death has been associated with either structural damage or physiological dysfunctions. The single practical criterion of death of microorganisms is the failure to reproduce in suitable environmental conditions.

Microbial inhibition The process of prohibiting, restraining, or hindering the growth of microorganisms, including the inhibition of enzyme activity within the organisms.

Microbial action spectra Plots or tabled values that describe spectral effectiveness of a photobiological or photochemical process of inactivation of an organism over a range of wavelengths.

Monochromatic light Light radiated from UV sources concentrated in only a very narrow wavelength range (bandwidth).

Optical power output (or radiant flux) The rate of flow of radiant energy from the UV light source per unit time and is expressed in watts (W).

Point-of-use (POU) Treatment devices designed to treat only a portion of the total flow of the product. In drinking water applications, POU devices treat only the water intended for direct consumption (drinking and cooking), typically at a single tap or limited number of taps. POU technologies are capable of removing microbial contaminants.

Photodynamic inactivation (PDI) The process based on the administration of a photosensitizer, which is preferentially accumulated in the microbial cells. The subsequent irradiation with visible light, in the presence of oxygen, specifically produces cell damages that inactivate the microorganisms. Two oxidative mechanisms can occur after photoactivation of the photosensitizer. In the type I photochemical reaction, the photosensitizer interacts with a biomolecule to produce free radicals, whereas in the type II mechanism, singlet molecular oxygen, O_2, is produced as the main species responsible for cell inactivation.

Photorepair processes: Molecular processes that are underlying the biological phenomenon of photoreactivation or photorecovery. Photoreactivation can be defined as the reduction in response of a biological system to UV light, resulting from a concomitant or posttreatment with nonionizing radiation. Although enzymatic photoreactivation is the most important phenomenon, other types of photoreactivation are also caught under this definition.

Plant photoresponse A response of a plant or other organism to light, mediated otherwise than through photosynthesis. The action spectrum for the overall photoresponse of plants, however, extends below 400 nm and beyond 700 nm, and there is a significant minimum in the green region.

Polychromatic light Light consisted of multiple wavelengths.

Reactive oxidative species (ROS) Highly reactive chemical species containing oxygen such as peroxides, superoxide, hydroxyl radical, singlet oxygen, and alpha-oxygen. In plants, the production of ROS is strongly influenced by stress factor responses. The factors that increase ROS production include drought, salinity, chilling, nutrient deficiency, metal toxicity, and UV light.

Sanitation The scientific discipline and practice of effecting healthy and hygienic conditions in production facilities. Sanitation uses an antimicrobial agent on objects, surfaces, or living tissue to reduce the number of disease-causing organisms to nonthreatening levels. Sanitizing does not affect some spores and viruses. Sanitizing a surface is meant to reduce, not kill, the occurrence and growth of bacteria, viruses, and fungi. Sanitizing should only be applied to food contact surfaces, which is required as part of the food code.

Shelf-life extension Uses methods of preservation that rely on applying chemical preservatives or physical agents and changing the storage conditions and/or the product packaging to inhibit microbial growth and increase shelf life of foods. A range of traditional methods for extending the shelf life of foods have been traditionally used including salt curing, smoking, pickling, refrigeration, freezing, and canning.

Ultraviolet (UV) light A band of the radiant energy of electromagnetic spectrum with wavelengths from 100 to 400 nm, shorter than that of visible light but longer than X-rays. UV light is present in sunlight and contributes about 10% of the total light output of the sun. Generally accepted ranges of wavelengths are based on physiological effects.

UVA light—long-wave, not absorbed by the ozone layer: soft UV, 320—400 nm. UVB light—medium-wave, mostly absorbed by the ozone layer: intermediate UV, 280—320 nm.

UVC—short-wave, germicidal, completely absorbed by the ozone layer and atmosphere: hard UV, 100—280 nm.

Wavelength Fundamental descriptor of electromagnetic energy including light measured in nanometers ($1 \text{ nm} = 10^{-9} \text{ m}$). It is a distance between corresponding points of propagated light.

Index